T0260488

# Dragons
# in the Dust

LIFE OF THE PAST
James O. Farlow, Editor

Ralph E. Molnar

# Dragons in the Dust

## The Paleobiology of the Giant Monitor Lizard *Megalania*

INDIANA
University Press
Bloomington & Indianapolis

Publication of this book was assisted by
the Friends of Indiana University Press.

This book is a publication of
Indiana University Press
601 North Morton Street
Bloomington, IN 47404-3797 USA
http://iupress.indiana.edu
*Telephone orders* 800-842-6796
*Fax orders* 812-855-7931
*Orders by e-mail* iuporder@indiana.edu

The paper used in this publication meets
the minimum requirements of American
National Standard for Information
Sciences—Permanence of Paper for
Printed Library Materials, ANSI
Z39.48-1984.

Manufactured in the United States of
America

**Library of Congress Cataloging-in-
Publication Data**

Molnar, Ralph E.
Dragons in the dust : the paleobiology
   of the giant monitor lizard Megalania /
   Ralph E. Molnar.
        p.     cm. — (Life of the past)
   Includes bibliographical references
and index.
   ISBN 0-253-34374-7 (cloth : alk.
paper)
       1. Megalania.  2. Paleobiology—
Southern Hemisphere.  3. Paleontology
—Pleistocene.  I. Title.  II. Series.
QE862.L2M65 2004
567.9'5—dc22          2003015569

1  2  3  4  5  09  08  07  06  05  04

It does not do to leave a dragon
out of your calculations, if you
live near him.

—J. R. R. Tolkien

# Contents

*Preface*

This is a book about one of the most extraordinary lizards that has ever existed. The giant monitor *Megalania* of Australia lived during the Pleistocene, popularly known as the Ice Age. It still existed when humans first came to Australia. Not only the lizard itself, but also the world in which it lived, are portrayed here.

We will begin in the first chapter with an exploration of the Pleistocene world: how long ago it was, what it was like, and how the physical environment—mostly the climate—affected the plants and animals. The impact of both "ice ages" (glaciations) and interglacials on living things and their evolution will be sketched out. We will look into how the changes in climate came about and consider how much environments in the Pleistocene may have been different and distinct from those that remain today.

Chapter 2 will look at the Pleistocene in Australia, starting with a brief history of evolution in Australia since the Early Cretaceous in order to place Pleistocene conditions in perspective. We will also look farther afield, to faunal evolution in Indonesia, and consider the history and biology of the ora *Varanus komodoensis*.

Returning to Australia in chapter 3, we will recount the history of the discovery of *Megalania,* with a brief history of fossil lizards in Australia. Chapter 4 will examine the relationships of *Megalania* to other monitors (varanids), both within and outside of Australia. Chapter 5 will be a reconstruction of the paleobiology of *Megalania,* followed in chapter 6 by a sketch of the factors that permitted and encouraged the evolution of *Megalania.*

# Acknowledgments

As always with writing a book, many people other than the author contribute much to the success of the endeavor. Laurie Beirne, Trevor Clifford, Barbara J. K. Molnar, and Ian Sobbe (all at, or affiliated with, the Queensland Museum), Richard Blob (Clemson College), Mike Lee (then at the University of Queensland), Frank Seebacher (now at the University of Sydney), and Steve Webb (Bond University) deserve special mention for their unfailing enthusiasm and assistance with this project. Ian Sobbe, Cecil and Doris Wilkinson, Malcolm Wilson, Robert Knezour, Russell Kneebone, J. Hawkins, and Henk Godthelp all collected and donated to the Queensland Museum specimens that played their part in making possible the new insights presented here. Greg Czechura and Steve Wilson (and Mike Lee) enhanced my understanding of living monitors. Joanne Wilkinson and Angela Marie Hatch graciously provided information and assistance with Australian fossil monitors, and Dr. G. J. M. Gerrits (Queensland Museum, Anthropology section) kindly translated material in Bahasa Indonesia on oras and fossil monitors in that country. Mike Archer, Jeanette Hope, Bob Jones, Anne Kemp, Randall Nydam, Tom Rich, Alec Ritchie, Steve Salisbury, and Jim Warren also contributed to the success of this project. I am also grateful to the editors at Indiana University Press for their criticisms and comments, which have improved the text.

This book and the research reported here were initiated at the request of Jim Farlow. Such merit as this work may possess is thus due in no small part to his request and to the inspiration of his research.

# Introduction:
# The Lizard of Aus

*Why dragons? The simple
answer to that question is
that dragons are very
satisfactory characters to
have around, provided it is
someone else who encoun-
ters them, not oneself.*

—R. WHITLOCK, 1983

The exhibit of *Megalania prisca* in the Queensland Museum rarely fails
to attract attention. A replica of a skeleton 5.7 meters (about 18 feet)
long makes an impact on even the seasoned museum visitor (Fig. I.1).
The animatronic restoration of the live animal reportedly led museum
guides to warn parents of young children, who often took fright at the
exhibit (Fig. I.2). Such nervousness was not limited to young children:
anyone who encountered live large Australian monitors (goannas) also
had sober thoughts about this beast (Fig. I.3). Even some professional
archeologists are of the opinion that no one could live long in a region
inhabited by these creatures.

The largest of all land-dwelling lizards appeals to our imaginations
in much same way as do dinosaurs. It reminds us of the myths and
legends of dragons, stories that not many centuries ago frightened our
ancestors. In its own limited Australian way, *Megalania* itself has
inspired such tales.

Late one night, so the story goes, a family on a remote farm in
northern New South Wales near the Queensland border was awakened
by the clucking of their chickens and barking of their dogs. Hastily
dressing and grabbing torches (Australian flashlights) to investigate,
they saw nothing outside in the yard, not even the dogs, which had
vanished into the nearby woodland in hot pursuit of whatever had
invaded the yard. By the next morning the dogs had not returned, and
they were never seen again. The chicken wire fence of the fowl yard had
been destroyed, and immense tracks led off into the woodland. The
family felt it best not to investigate any further. The tracks were those
of a gigantic lizard (Gilroy 1995).

Does this sound like a scene from some down-under version of *The
X-Files*? Here's another. At its eastern end, the New South Wales–
Queensland border extends through very rough country. Low volcanic
mountains are cut by steep, narrow ravines and overgrown with one
of Australia's last remaining rain forests. It is dim within the forest, with
a thick canopy overhead. Although there is little underbrush, one cannot see very far because of the roughness of the terrain. Bird calls can
be heard high overhead in the canopy, and small, nondescript brown

birds—locally known as "housewife birds"—hop along the forest floor, tossing small leaves helter-skelter into the air in search of tasty insects beneath. Although much of the forest is a national park, there has been logging in adjacent areas. In 1955 a timber cutter walking through the forest encountered a giant lizard. The head of the creature shot forward and—seizing his hand—devoured two fingers. One of his fellow loggers sprang forth and plunged a railway spike into the head of the beast, killing it instantly. Measuring the carcass before leaving, presumably to get medical attention for the victim, they found it to be fully 20 feet long. When their mates returned a few days later, the carcass was gone (Gilroy 1995).

Although there are a number of large monitors in Australia—called goannas—they don't reach a length of 10 feet, much less 20. So these are claimed to be sightings of survivors from the past of the giant prehistoric monitor *Megalania*. Cryptozoologists—those who seek to track down new and unknown animals like the yeti, sasquatch, or Loch Ness monster based on reports like these—have gathered almost thirty such reports. But so far, they have only reports. And not all crypto-zoologists take these reports seriously. Many of them are doubtless basically honest accounts of encounters with large monitors. Australia has several species, the largest of which reaches two meters (about 6 feet 6 inches) long. When, as we shall see later, even professional zoologists tend to overestimate the length of the large monitors, the oras of Komodo Island in Indonesia, it isn't surprising that other people do too.

*Fig. I.1. (opposite, top) Replica of the skeleton of* Megalania prisca, *made under the direction of Dr. Tom Rich at the Museum of Victoria, Melbourne. This represents an individual 5.7 meters long, approximately the largest estimated length. Copies of this reconstructed skeleton were distributed to museums throughout Australia, and this one was in the Queensland Museum, Brisbane, in 1990.*

*Fig. I.2. (opposite, bottom) The animatronic model of* Megalania prisca *in the Queensland Museum (in 1990). The creature is represented as standing at the bank of a gully.*

*Fig. I.3. A large Gould's goanna,* Varanus gouldii, *seen lurking along the Burketown road in northwestern Queensland in June 1977. Remains of* Megalania *were collected on the same trip.*

When a lizard is seen briefly, or only in part, accurately estimating the length requires more skill than most people realize.

I find it interesting that none of these reports predates the discovery of *Megalania,* which was announced to the scientific world in 1859. In fact, Australian paleontologists were well aware of the beast by the 1880s. Most of the reports of encounters with giant lizards came after the Second World War, when increased educational levels brought this animal to the attention of the public. Also significant is the recent discovery that *Megalania* had a crest on the top of its head, something lacking in all modern monitors. None of the alleged *Megalania* sightings mentions what would have been a prominent feature—probably of the males.

Although certain cryptozoologists claim there are abundant references to *Megalania* in aboriginal mythology, almost all of these seem to be simply references to lizards. The only example of a plausible reference to *Megalania* that I could find came from the Murray River region along the Victoria–New South Wales border. This was the story of a giant goanna, the whowie, recounted by Smith (1930). According to the legend, the creature was about 20 feet long but had six legs rather than the usual four. It certainly isn't impossible that stories about *Megalania* may have survived among the aborigines. But working out which stories referred to *Megalania* and which were inspired by modern goannas may be another matter. We shall look no further at such stories but will look instead at what can be learned of the beast itself and the world in which it lived.

One significant issue—Pleistocene extinctions—deserves some mention here. This topic has caused much discussion, and this is not the place to treat it thoroughly, since we will be more interested in how *Megalania* lived. Nonetheless, since *Megalania* is extinct, some brief mention is relevant.

The Pleistocene extinctions at the end of the last ice age are usually attributed to either of two causes: climatic change (of which, as we shall see, there was no lack), or human hunting and other interference (of which there was also no lack). Both are obscured—especially in Australia—by emotional issues, particularly a love for nature and a Rousseau-like reverence for "traditional" (nonindustrial) peoples.

The hypothesis of climatic change faces the problem of explaining why, when apparently similar climatic changes have recurred over the past 2 million or more years, it just happened to be the one that occurred when people were around that caused the extinctions. So far, this has not been adequately explained, and the dating of the extinctions in Australia to a time when there was no obvious climatic change (Miller et al. 1999) does not help the hypothesis of climatically caused extinction. Evidence published just as this work was finished (Roberts et al. 2001) is consistent with an extinction caused—in some fashion—by the arrival of humans and not at a time of significant climatic change.

However, we should not be surprised if the causes of these recent Pleistocene and post-Pleistocene extinctions are not monolithic if different ones occurred in different places. There is little doubt (among scientists, if not in other quarters of society) that the extinctions on the

Pacific islands were the result of human hunting or other interference. This seems increasingly likely for Australia as well, although whether it applies to the Americas—for which the hypothesis was originally proposed—is less clear. On the other hand, the bulk of the evidence suggests that the extinctions in northern Siberia, the Bering land bridge, and Alaska were due entirely to natural causes. This evidence (discussed in the next chapter) suggests that the environment in which these creatures lived, the "mammoth steppe," has disappeared entirely from the earth.

So just 50,000 years ago the world was filled with fabulous creatures that are now forever gone. There were woolly elephants and rhinos; there were saber-toothed cats, great wolves, and lions; there were giant rhinos, each with a single massive horn on the forehead; there were elk with wide racks of antlers, and others with antlers in the form of vertical plates; there were giant tortoises and giant monitors. That such creatures of legend are now gone—aside from being a minor relief for some—is an indication of how much the world can change and has changed with time. How this happened, which factors change with time and which do not, is a major concern of paleontology and historical geology.

In concentrating on *Megalania* we will face these issues, for this largest of all land-dwelling lizards raises many questions. How did it evolve? What did it eat? But perhaps most interesting of all, how—in a time when mammalian predators dominated other continents and even the seas—could a lizard ever come to be so large and powerful?

Like visiting a great mountain, we will approach our questions from a distance, to see them rising slowly over the horizons of our understanding. In this way the perspective necessary to answer our questions—and to appreciate the answers—is gained. So to understand *Megalania,* we must also understand its world, the world of the Pleistocene—from 1,800,000 until approximately 10,000 years ago (Gradstein and Ogg 1996).

You will encounter phrases like "seems to be," "plausibly," "presumably," and "probably" many times in the text of this book. What we know of the life of the past, although indisputably exciting and deeply informative, is often very limited. It is difficult today to obtain accurate information on the lives of many living animals. For most extinct forms this is impossible. Although I would be willing to wager money, sometimes at long odds, that many of the conclusions reached here about *Megalania* are correct, we must always remember that the conclusions are only as good as the data on which they are based. And the data are not always as good as we might desire.

It seems almost compulsory in introductions to have some rather dry comments about conventions to follow in the text of the book. These will be minimized, but there are a few points that should be mentioned here. (Anyone familiar with this material can skip ahead to the next chapter.) One point has to do with biological (taxonomic) names. All known species, living and extinct, have a biological name. These were originally, and still often are, in Latin because Latin was spoken by all educated Europeans in the eighteenth century, when taxonomic names were first widely applied. The point of this was to ensure that each organism had one, and only one, name (of two words) and that this name was the same in all languages. This was, in fact, a rather naively optimistic hope, but by and large it has been retrieved by having only a single valid name. More than one name may be assigned to one organism by mistake—the scientist involved being either unaware that the organism has already been named or else believing (mistakenly) that it is a new and distinct form. This does not play much role in understanding *Megalania*, since it is rather difficult to mistake such a large monitor for anything else; however, as we shall see, it was done.

More relevant is the policy of basing names on evolving lineages. What we might call the minimum lineage is the species. (Actually there may be even smaller groups of evolving lineages, but they are generally not named.) Species are defined as being composed of a group of interbreeding or potentially interbreeding individuals. They are given a name of two words, such as *Megalania prisca*. The first word, *Megalania*, is also the name of a more inclusive group, the *genus*. The name of a species is often abbreviated by citing only the first letter of the generic name together with the specific name, for example, *M. prisca*. By itself *prisca* isn't really a name—although it is termed the "trivial name"—it is just a word (in Latin). As in the genus *Varanus*, there may be more than one species assigned to a genus. The genus *Varanus* also has a vernacular name, "monitor" in English (there are others in other languages). In Australia, it is traditionally called the "goanna," which derives from "iguana" (which it isn't). Just as one or more species may group together to form a genus, so one or more genera may group together to form a subfamily, family, or other more inclusive category. For monitors, the familial name is Varanidae, and the subfamilial name Varaninae, often given as "varanids" or "varanines." These names indicate increasing degrees of relatedness, from family through subfamily, then genus, to species. "Varanids" effectively means "monitors."

Other esoteric abbreviations used in the names of animals (and plants) include "sp.," "cf." and "aff." The first of these, "sp." is used when a specimen meets the criteria to be assigned to some genus but is too incomplete or poorly preserved for the species to be determined. The other two are also applied to problematic specimens: "aff." means that the specimen is thought to be related to the species whose name follows the abbreviation. And "cf." (an abbreviation of a term literally meaning "to compare") is usually used to indicate similarity. Thus *Varanus* aff. *Varanus bengalensis* is a monitor thought to be related to *Varanus bengalensis*, and *Varanus* cf. *V. bengalensis* is one that is similar to *Varanus bengalensis*.

In chapter 5 we will be discussing specific individuals of *Megalania* and their individual bones. These bones are held in museum (or other) collections, and a specimen number (inked directly on the fossil) identifies each specimen and links it to the information about that specimen recorded in the museum's catalogs (or registers). Each number is also recorded on a specimen label kept with the specimen in the collections or on display and in the catalog itself. In principle, each number refers to an individual animal, but in the case of scattered bones it is not always possible to be certain that all of the bones derive from a single individual. Thus, different bones from the same fossil site (perhaps even the same individual) may be given different numbers; whereas if they were found articulated as at death, or if there is other good reason to conclude that they all derive from one individual, then they all receive the same number. These catalog numbers usually consist of two parts: first a series of letters that designates the institution, followed by the number itself (which may also include one or more letters in addition to numerals). The series of letters is often (but not always) an acronym of the name of the museum or other institution, such as AMNH for the American Museum of Natural History in New York, AM for the Australian Museum in Sydney, or QM for the Queensland Museum in Brisbane. The terms used in this book are given as "collection abbreviations" at the end of this introduction. The number provides a unique "label" for each specimen—at least in theory; sometimes mistakes occur. Thus, we will encounter numbers like QM F873, which indicates Queensland Museum fossil specimen (hence the "F") eight hundred and seventy-three.

In giving measurements, the metric system is usually used, except when another work (or person) is cited that used the imperial system. In that case, the measurements are given in imperial units.

In discussing monitors, we inevitably must mention monitors other than *Megalania*. For these the accepted vernacular names (given, for example, in D. Bennett 1998, or Steel 1997) will generally be used. The lizards often called "Komodo dragons" are called oras, the native Komodo islanders' name for them. This follows the usage of Walter Auffenberg (1981) and emphasizes that oras are giant monitors, whereas dragons are mythological creatures that perhaps were inspired by oras.

## Collection Abbreviations

AM      Australian Museum, Sydney

QM      Queensland Museum, Brisbane

UCMP    University of California Museum of Paleontology, Berkeley

# Dragons
# in the Dust

# 1. The Pleistocene World

### Introduction: Back to the Ice Age

*Megalania* lived during the Pleistocene, the "Ice Age." Like other organisms, it evolved or faced extinction according to the dictates of the environment. Thus, understanding *Megalania* presupposes understanding the Pleistocene world. Furthermore, since the Pleistocene was the mother of the present, even to comprehend our own world we must know the Pleistocene. For in the Pleistocene, the present was born.

I once briefly lived in Fargo, North Dakota, a part of the North American continent so flat that people used freeway off-ramps for downhill skiing and traveled for 30 minutes to the north, toward Grand Forks, just to see a hill. And it was only about a meter high. To those of us who came from the Rockies or the Appalachians, this country was amazingly flat. One could not help but wonder how it came to be so. To the east, in Minnesota, there are odd, rounded, little gravelly hills—kames—sitting on the otherwise flat and rather monotonous plain. There also, a bench-like terrace in a hillside could be found; it marked an old Pleistocene beach—not a beach from the ocean, but from a vast lake, now drained and dry. The region around Fargo was flat because it had been the floor of Lake Agassiz, a great but (geologically) short-lived body of water that at its maximum extended from near Minneapolis, Minnesota, to near Reindeer Lake on the Manitoba-Saskatchewan border. Lakes Winnipeg, Winnipegosis, and Manitoba are probably its remnants. The lake's water came from the melting of the great ice sheet. The kames were deposits laid down on the sheet of ice that became hills when the ice melted and left them behind. Relics of the great glacier not only littered, but actually formed, the entire landscape.

It is the evidence for these glaciers of the Pleistocene that gave it the name "Ice Age." But this is just the most recent example of such an episode in the history of the earth. During this one, much of the landscape of North America and Europe was overrun by glaciers, as was Antarctica in the south, where the ice still remains. There have been several major advances and retreats of the ice—glaciations of the continents and seas—during the past two million years. Four of these occurred during the Pleistocene, separated by warmer periods, the *interglacials,* and punctuated by short transient warmings, the *interstadials.*

These different periods required names in order to be clearly kept in mind. And names required definitions, ways of distinguishing one time from another. Periods of time are divided, or defined, by events: the positions of hands or changing digits for clocks; significant and obvious (to geologists) events for geological periods. For the Pleistocene, the obvious events are the sequence of advances and retreats of the glaciers. Different names have been given to the glacial periods in different places (Table 1), because at the beginning of our understanding of the Pleistocene no one knew which periods at any one location corresponded to those elsewhere. It wasn't even known if different places had experienced the same number of advances and retreats of the ice (and hence comparable periods).

There are no names in general use for these periods in Australia, so—arbitrarily—the names for Alpine Europe are used here for the glacials, and those for northwestern (continental) Europe for the interglacials. When the Pleistocene is considered over the whole world, geologists generally use either the Alpine or North American terms. Although the correlations shown in Table 1 are generally accepted, these correlations are still not as well established as we could wish, especially regarding the *pluvials* (a term originally thought to denote periods of greater rainfall) of Asia and Africa. The pluvials were initially believed to have been contemporaneous with the glacials but are now correlated with interglacials: aridity characterized the glacials and apparently was especially intense during their opening and closing phases. Dust in glacial and sea-floor sediments is taken to indicate dry, windy conditions—which blew the dust onto the glacier or into the sea—and to correlate with glacial times. The ages are established by radiometric dating, that is, dating using radioactive decay—for example, the carbon 14 or potassium/argon method.

The discovery that some rocks are magnetized allowed a different and independent set of criteria to be used to correlate sequences of rocks and sediments: their magnetization. The magnetization of a rock follows and preserves the direction of the terrestrial magnetic field at the time the molten rock cooled and solidified. If small sedimentary particles are magnetized and if they contain iron, they may also be laid down so as to record the direction of the magnetic field. Thus, igneous rocks as well as some sedimentary rocks can be used to determine the directions of ancient fields. If the terrestrial field never changed, then determining the magnetization of ancient rocks would only tell us what we already knew; but in fact, the field does change from time to time. Sometimes, as now, the south magnetic pole was directed toward the north geographic pole (called "normal polarity"); at other times, toward the south pole ("reversed polarity"). The south magnetic pole is at the north geographic pole because the north pole of a compass needle was defined as the end that pointed northward. The north pole of one magnet is attracted to the south pole of another, and since the north

## Table 1
### Names and ages of the glacial and interglacial periods

| Alpine Glacials (and Interglacials)* | North American Glacials (and Interglacials)* | Northwest European Glacials (and Interglacials)* | Russian Glacials (and Interglacials)* | Approximate Ages (my = millions of years; ky = thousands of years) |
|---|---|---|---|---|
| Biber | "Nebraskan" | Pretiglian | not represented | 2.4 my |
| (Biber/Donau) | not represented? | (Tiglian) | not represented | |
| Donau | "Nebraskan" | Eburonian | not represented | 1.8 my |
| (Donau/Günz) | not represented? | (Waalian) | (Kryzhanov) | |
| Günz | "Nebraskan" | Menapian | Oka/Demyanka | |
| (Günz/Mindel) | (Aftonian) | (Cromerian) | (Likhvin) | |
| Mindel | Kansan | Elster | Dnepr/ Samarovo | 450-420 ky |
| (Mindel/Riss) | (Yarmouth) | (Holsteinian) | (Odintsovo) | 420-360 ky |
| Riss | Illinoian | Saale | Moscovian | 200-130 ky |
| (Riss/Würm) | (Sangamonian) | (Eemian) | (Mikulino) | 130-120 ky |
| Würm | Wisconsin | Weichsel | Valdayan/ Zyryanka | 115-10 ky |

* Interglacials given in parentheses.

pole of the needle is attracted toward the north geographic pole, that pole must be associated with the south magnetic pole. The direction of the field changes abruptly, geologically speaking, so these changes provide convenient "datum planes" (or "time data"), records of events occurring around the world almost instantaneously that can be used to correlate the ages of rocks in different parts of the world. "Almost instantaneously" geologically speaking, that is. The two magnetic divisions of the Pleistocene are the Matuyama (reversed polarity) and, since 730,000 years ago, the Brunhes (normal polarity).

Just as the glacial episodes had short interruptions, the interstadials, so (sometimes) did the magnetic epochs. For example, the Brunhes was interrupted by the Jaramillo (reversed) from 940,000 to 880,000 years ago. Magnetostratigraphy, the study of the temporal relationships of rocks based on their magnetization, has proved very helpful in the establishment of ages and correlation of rocks.

The temperatures of the water of the sea's surface changed with the climatic changes of the glacials and interglacials. These temperatures were reflected both in the kinds of organisms living in these waters and in the chemical processes occurring there. Since the shells of the organisms sink when the organisms die, they all collect in the sediments of the sea floor. Thus, these sediments record a history of the temperatures of the sea's surficial waters and of the air over them. Paleoclimatologists may use either the kinds of organisms or the composition of their shells (chemical processes of their construction) to infer past climates. In practice, both methods often make use of the shells of foraminifera (forams): those forams that lived in warm waters could be recognized and distinguished from those that lived in cold. David B. Ericson at Lamont Geological Observatory in New York (now the Lamont-Doherty Earth Observatory) pioneered reconstructing the history of the climate by tracing the sequence of warm- and cold-preferring forams. The proportion of two isotopes of oxygen, $O^{16}$ and $O^{18}$, that were incorporated into the tests (shells) of forams depends on the water temperature, so this ratio could be used to reconstruct ancient temperatures and thus climatic histories. This work was undertaken by Cesare Emiliani. Both methods worked, but it took some time to iron out the apparent discrepancies between them (see Ericson and Wollin 1964, and Imbrie and Imbrie 1979). The two methods revealed the same history and, after some effort, were shown to correlate with climatic history as preserved in the continental deposits as well.

There had been glaciations before the Pleistocene; in fact, cool and warm times seem to have alternated throughout the history of the earth. This pattern occurs at various scales—the cool periods lasting for years, millennia, or millions of years—and it extends as far back as we can tell. There were extensive glaciations during the Precambrian—at least four of them—and a restricted glaciation near the South Pole during the Ordovician (about 445–495 million years ago). The Precambrian, that period of time from the beginning of the earth to (more or less) the appearance of multicellular life, was obviously long—around 90 percent of the history of the earth—so it is not surprising that it has a number of glaciations. What is significant is that the glaciations started early. One of these glaciations (the Varangian) seems to have been the greatest ever and may have extended into the equatorial regions, bringing glaciers about as close to the equator at that time (about 1,100 km or 680 miles) as Trinidad is today (Evans, Beukes, and Kirschvink 1997). There is extensive evidence for Permo-Carboniferous glaciations in South America, India, southern Africa, and Australia, which also convinced geologists (initially those of the Southern Hemisphere) that the continents had been arranged differently then than they are now.

During the Pleistocene glacial periods the climate was cooler than it is now—from about 5°C lower near the equator to about 15°C lower at high latitudes—and glacial ice sheets spread over the northern continents and Antarctica. Interestingly, one can think of this as a depression of the layers of the atmosphere: the climate of the ice age still exists about 2 km above our heads at sea level (Galloway 1965). It was the evidence for the ice sheets that also indicated that the climate had been cooler. Two hundred years ago most geologists did not realize that there had been any ice ages at all, much less a recent one. Swiss peasants, who were familiar with evidence of montane glaciers, recognized that the Alpine glaciers had once been much more extensive than they are now; whereas geologists still interpreted this evidence in terms of a worldwide deluge (Imbrie and Imbrie 1979). The community of geologists became convinced by a suite of features, each of which could most convincingly be explained as the results of glaciers where none now exist. When taken together, this evidence indicated unexpectedly extensive glaciation.

These features included moraines—elongate hills of rocks and rock fragments—deposited by glaciers, as well as the shorelines of certain lakes, like the Great Salt Lake, that are now well above the lake's surface. Valleys with U-shaped cross sections and amphitheater- or bowl-like valleys often found high in mountains, known as cirques, both testify to the erosive power of glaciers. So do fluted (grooved) surfaces developed in rock or sediment, and polished and scored surfaces of exposed bedrock, both caused by the abrasive action of moving ice. Drumlins—elongate, smoothly rounded hills, shaped like inverted bowls of spoons—resulted from the flow of ice over preexisting hills; and eskers—long, narrow, often sinuous ridges—proved to be deposits from glacial streams. However, perhaps the most influential features (at least in accounts of the history of how the glacial ages came to recognized) were erratics, boulders of rock unlike that found anywhere in the vicinity. Erratics posed the questions Where did they come from? and How did they get there? Initially erratics were thought to be relics of the biblical flood, but careful study made it clear that they had been moved by ice.

Sedimentary as well as geomorphological features resulted from glaciation. *Drift*—masses of rock and rock fragments—and *loess*—very fine dust often deposited in thick masses—are prominent among them. The ice sheets left a clear geological sign in the form of *till*. An undisputed relic of glaciers, till is composed of unsorted rocks, cobbles, gravel, sand, and mud all mixed together, not in orderly beds with the larger pieces toward the bottom and the smaller toward the top. Till is the result of detritus carried by glaciers and deposited in place by their melting.

## Physical Environment of the Ice Age (Glacials)

At its greatest extent, the ice covered much of northern North America and Europe and the entire Arctic Ocean. The ice sheet did not extend out from a single source in the northern continents. It does so now in Greenland only because Greenland is a relatively small island (compared to a continent). In North America three separate sheets coalesced. The Laurentide sheet, centered on what is now Hudson Bay, was the largest and merged with the Cordilleran sheet centered on the Canadian Rockies in the west and the Greenland sheet in the east. Three major sheets, none as large as the Laurentide, also made up the Eurasian glaciers. The Fennoscandian sheet, centered on Scandinavia, was the largest and joined with the British sheet in the west, centered on the northern British Isles, and the Barents sheet, centered on the Barents Sea, in the east. The Fennoscandian sheet also merged with the Greenland sheet to the northwest (Fig. 1.1).

Recent evidence suggests that not only the adjacent land but also the Arctic Ocean was completely frozen over, with perhaps as much as a kilometer of ice covering it (Polyak et al. 2001). The thickness of the continental ice sheets is unknown, but it is generally considered to have been about 3.5 km (about 2.2 miles) for the Laurentide. Dissenting opinion suggests it was only a little over 2 km (Kerr 1994). Geologists had generally assumed that the cooler climates of the glacial times implied more precipitation, whether rain or snow. But the removal of water from the oceans and its storage as ice should have created drier climates since less water was available to evaporate because of the reduced size of the oceans. Furthermore, that water was cooler, thus requiring more energy (i.e., heat) to evaporate.

Work during the 1970s had suggested that changes in temperature during the glacials occurred preferentially in the polar regions. In other words, it was not that the whole planetary surface (which here effectively means the surface of the ocean) had cooled as such, but that the cooler (i.e., polar) regions had gotten even colder, while the equatorial regions remained more or less as warm as they are now. More recent evidence disputes this and suggests that both polar and equatorial regions cooled—by about 5°C in the Tropics (Charles 1997), although the possibility remains that some equatorial regions didn't cool significantly.

The glacial ice sheets stored not only water but also—speaking metaphorically—cold. They chilled the overlying air. Cold air is denser than warmer air and thus heavier. This alone would cause it to flow down the ice sheets, as what meteorologists term *katabatic winds*. Such winds occur today in Antarctica and Greenland and may blow at more than 150 km/hr (about 93 miles/hr). These strong winds also, presumably, blew off the Pleistocene ice sheets of Europe and North America and may have been important in forming the extensive loess deposits of those lands. They were clearly cold, although it has been argued that

Fig. 1.1. The North American and European ice sheets of the Late Pleistocene at their maximum extent. Mountain glaciers, mostly in Asia, are also indicated. We look directly down at the North Pole, with the Greenwich Meridian extending vertically from the bottom (not drawn on the map). Heavy lines show the areas of the ice sheets and glaciers, and light lines show the shorelines (dotted lines indicate shorelines overrun by the ice sheets). Note that the shorelines shown are modern, not those contemporaneous with the ice sheets. Key: 1. British icesheet; 2. Fennoscandian sheet; 3. Barents sheet; 4. Greenland sheet; 5. Laurentide sheet; 6. Cordilleran sheet. The three components of the European ice sheet can be readily seen, but the Cordilleran sheet is broadly united with the Laurentide: their boundary extends southeastward, not far from the Pacific coast, from the indentation in the margin of the ice sheet in Alaska. (From Kurtén 1972.)

because of heating resulting from the increase in density of the air as it descends (adiabatic heating), these winds were not substantially colder than winter winds in these regions today. Judging from the winter in North Dakota, that is quite cold enough.

The ice sheets affected the general circulation of the atmosphere in the Northern Hemisphere. Simulations indicate that the jet streams over North America would have been split—the southern one bringing increased rain to the southwest and so presumably accounting for the now-dry lakes there (Fig. 1.2). The Laurentide ice sheet also affected the winds blowing out across the Atlantic, causing cooler and stronger westerlies.

The interglacials, on the other hand, were perhaps even warmer than the present climate. Evidence for this can be seen in the distribution of mammals in Europe, where hippos waded in the Thames and elephants (real elephants, not mammoths) walked the Downs of southern England.

The water for all the glacial ice must have come from somewhere. Most of it derived from the great reservoirs of water on earth, the oceans. The degree to which the sea level fell is debated. It may have been by as much as 200 meters (about 655 feet), but it was at least by about 130 meters (about 425 feet). As sea levels fell, the areas of the continents expanded. Thus, during the Würm there were differences in geography from today (Fig. 1.3). There were no British Isles; instead,

Fig. 1.2. North America during the maximum extent of the ice sheets of the Würm (Late Pleistocene). The ice can be seen in the north, and a series of great lakes—now dry—in the far west (the Great Lakes, in the east, had yet to be formed). The shorelines shown are contemporaneous coasts, not modern ones. (From Lamb and Sington 1998.)

Fig. 1.3. The world during the Würm. Contemporaneous shorelines are shown in heavy lines, moderns coasts by dotted lines. The ice sheets are not shown (see Figs. 1.1, 1.2). This geography is different, yet almost familiar. "B" indicates Beringia. Note that in this projection, areas near the poles are exaggerated. (From Chorlton 1983, and Lamb and Sington 1998.)

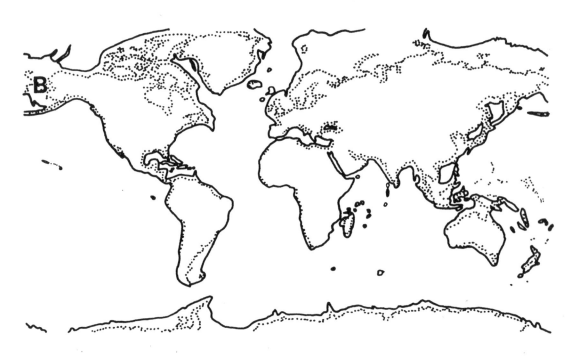

they were part of the European mainland. But there were large North Atlantic islands that are now submerged, reminiscent of those—like Buss—of mariners' tales of five centuries ago. Extensive islands in the Arctic Ocean were not completely glaciated, at least not all of the time. A land bridge, Beringia, linked Alaska with Siberia. An island approximately the size of Cuba stretched where the Bahamas are now, and the Falklands were a contiguous part of Argentina. Japan was part of the Asian mainland, with what is now the Japan Sea an inland sea between it and China. New Guinea was continuous with Australia, and Fiji in the South Pacific and the Andamans and the now-submerged Mascarene Plateau in the Indian Ocean were islands about as large as Sardinia. Only Antarctica and Australia (with New Guinea) remained isolated continents: one could, climate and fauna permitting, walk from the Cape of Good Hope to Cape Horn. Similar changes in geography must have occurred with each of the glacials, the fall in sea level being more or less severe depending on how extensive the ice sheets were. The fall was not of an on-or-off nature, however. During the last glacial, the sea level was about 50 meters below the modern level for almost 90 percent of the time, but it did fall to at least 130 meters below toward the end, about 20,000 years ago.

The changes in sea level would have increased the variability of the climate, both because continents tend to heat more and faster in the summer and cool more and faster during the winter than the oceans and because the continents were effectively made higher. It has even been suggested that the lowered sea levels allowed the release of large quantities of methane that had accumulated in sea-bottom sediments (about which more later), thus adding a greenhouse gas to the atmosphere and so helping to reverse the conditions that brought about the glacial. The lowered sea level, combined with an Arctic Ocean completely frozen over, would have left Beringia and northwestern Siberia without nearby sources of moisture (Fig. 1.4). Thus, this region would have been drier than it is now.

Of course, during the interglacials the sea level rose, and although the geography in general may have been comparable to that of today, there were some differences. At one time, for example, the level of a predecessor of the North Sea (known as the Holstein Sea) was high enough for the Jutland Peninsula of Denmark to be an island. Since the late 1980s work on shoreline subsidence has increasingly suggested that sometimes, and in some places, the sea level rose rapidly (e.g., Blanchon and Shaw 1995), by as much as 7 cm/year (cited in Tooley 1989). These rises were "fueled" by the melting of the ice sheets and the draining of the adjacent (proglacial) lakes into the sea. Rises in sea level had effects on marine communities as well as on those of the land. Coral reefs—shallow water communities dependent on sunlight—that became established in during glacial times could be "starved" of sunlight as the interglacial sea level rose too quickly for them to match with their upward growth. An extensive barrier reef along the southwest coast of India, now below the level of coral growth, was formed during the last glacial and its coral killed off by the subsequent rise of sea level.

*Fig. 1.4. Siberia (S) during the Würm. Heavy lines indicate contemporaneous coastlines or, where edge-lined, ice sheets and the ice cover of the Arctic Ocean (A); light lines show modern coastlines; lakes are indicated by vertical hatching. Much of what is now the continental shelf (C) of the Arctic Ocean was then extensive dry tundra. (This projection does not exaggerate areas near the pole.) Although there were some montane glaciers (M), much of Siberia was distant from either the Barents ice sheet (B) or the open waters of the Pacific (P) in the southwest, Lake Baikal (L) in the south or the proglacial lakes (G) along the margins of the ice sheet to the west. The dotted lines in the eastern part of the ice sheet indicate a possibly ice-free corridor. (From Velichko et al. 1984.)*

This perspective of rising and falling seas and expanding and contracting lands tells us that a knowledge of modern geography is not enough to understand the distributions of animals and plants. We need to know not only how geography is but also how it was. We can see from the map of the world during the glacial times that lands now separate were then joined and that lands now submerged were then emergent. The junction of Asia with North America by the Bering land bridge explains many of the similarities in their faunae. The now-vanished Atlantic lowlands of the northeastern United States and south-eastern Canada provided refuges from the ice caps for both plants and animals of eastern North America and so help to explain the modern distributions of both animals and plants in that region.

Ice might not seem to be particularly heavy, but when it has accumulated to thicknesses of a kilometer or more, we can appreciate its weight. The land sinks under so much ice. This resulted in the slow depression of the regions under the ice sheets, especially in northeastern North America and northern Europe (Scandinavia). With the melting of the ice, the land rose again. These regions are now "rebounding," and the seas—Hudson Bay and the Baltic—that had crept over them are draining back into the Atlantic. This draining, although not quick enough to watch, is dramatically shown by the wreck of a wooden ship that was once on the shore of Hudson Bay but now is almost a kilometer inland (illustrated in Bray 1962, 46). The rises and subsidences are usually believed to have been slow. But there is evidence for substantial faults in northern Scandinavia; hence, the melting of the ice

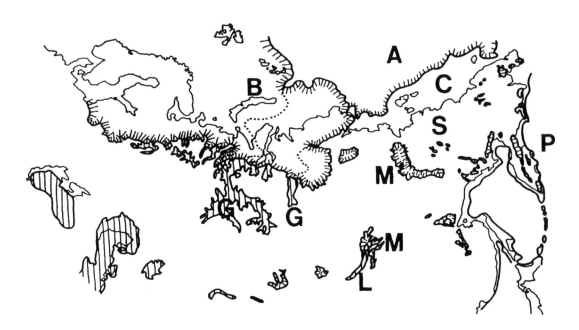

sheets may have been accompanied by more or less severe earthquakes (Arvidsson 1996).

With the withdrawal of water from the oceans, shallow seas—the Black Sea in particular—became isolated and began to dry. But when the ice sheets began to melt, some regions that are now dry land not near any oceans but well within the increased continental areas were submerged. The Caspian expanded into an inland sea in Asia of more than twice its current area at the end of the Würm, and there is evidence for intermittent, more extensive "seas" there (as well as in North America) during the interglacials. Proglacial lakes—lakes adjacent to the glaciers—were formed by meltwater from the ice sheets collecting at the glacial edges, where the continental surface was depressed by the weight of the ice. In Asia there was a second cause of now-vanished lakes: the Barents ice sheet blocking the course of the northward-flowing rivers such as the Ob. These also formed lakes along the margins of the ice sheet, which drained to the Aral and Caspian Seas to the south.

The edges of the ice sheets were an environment unlike any now existing. For one thing, the edges were—in the Northern Hemisphere—well south of the modern ice sheet in Greenland. They were therefore exposed to more and stronger sunlight than such glaciers today. The katabatic winds would have kept the margins cold and dusty. Examination of lake sediments from Maine and New Brunswick indicates that there were strong temperature gradients near the edge of the ice, possibly as much as now occur over 1,500 km (about 930 miles) of latitude (Levesque, Cwynar, and Walker 1997). The cold air over the ice sheet would have created a stationary front, with warmer continental air flowing up over the cold air of the ice sheet. As in other fronts, the rising air would cool, and its moisture would condense as cloud from which rain would fall as cooling continued. It is believed that the cloud would form as stratus and fog along the margin of the ice and that this would reflect solar heat and keep the region of the margin cold. But a mammoth wandering away from the sheet would come out from under the cloud into a region heated by the sun to much the same temperature that it is now. This combination of "storing" cold and reflecting sunlight maintained the strong temperature gradient.

## Biological Effects of the Ice Age (Glacials)

Among the ecological effects of the glaciations was "telescoping" the habitat for plants (and animals), reducing the distance between the frigid glacial faces and the tropics. It was about 1,900 km from the Tropic of Cancer to the ice sheet at the last glacial maximum, compared with 4,100 km to the (Greenland) ice sheet now. And this was repeated each time the ice sheets advanced. Refuges—regions where the environment remained unchanged—were created. Some refuges were near-glacial lands where the ice sheets happened not to fully cover the land surface, as in mountain ranges and coastal lowlands. Others, further south, were where local conditions permitted survival of the ecosystems of interglacial—or preglacial—times. The role of refuges appears to have been more important in North America, where the mountain chains are oriented north-south, thereby permitting southward and northward movement, than in Europe, where the chains are oriented east-west and are "reinforced" by the similarly oriented Mediterranean Sea, thus inhibiting north-south movements. The importance of refuges is shown by Antarctica, which has none, and hence where the land-dwelling fauna has been almost entirely exterminated by spreading ice sheets. When animals (or plants) repeatedly invaded (from refuges) territory uninhabited because it had been covered by glaciers, their genetic variability may have been reduced. Computer simulation indicates that the populations expanding into the new areas come largely from a small subset of the populations in the refuge and that during the process of expansion, genes (alleles actually) are lost (Hewitt 1996). We must also remember that just as the more temperate southern regions provided refuges during the glaciations, southern mountains provided refuges during the interglacials for animals and plants adapted to the cold, as they do now.

Pollen data collected about fifteen years ago indicates that the tropical regions in southern North America and northern South America that now support rain forests supported more open vegetation during the Late Pleistocene (Lewin 1984). It is still unclear whether and how much this affected the Amazonian rain forest—some claim it was reduced to moister "pockets"; others suggest it changed more in composition than in extent—but the tropical African rain forest seems to have been reduced and in parts replaced by dry scrub (Moore 1998). This suggests that the "ancient" rain forests of these regions (and perhaps others) are not really ancient but were formed anew with each interglacial, like the Great Barrier Reef of Australia, as we shall see in the next chapter.

## Conditions During the Interglacials

During the interglacials conditions would have "returned" to those similar to today's. The climate ameliorated, and the sea level rose. For populations of land-dwelling organisms, isolation would have increased, and hence the opportunities would also have increased for evolution to take different courses in response to different local conditions or just by chance.

When the glaciers melted, the meltwater literally as well as figuratively went somewhere, often in torrents. There is evidence for catastrophic flooding in South America, in the Rio Beni region of Bolivia (Campbell, Frailey, and Arellano 1985). Thick deposits of conglomerate about 10,000 years old and a linear chain of lakes are attributed to the catastrophic overflow of Lake Titicaca, swollen by water from melting Andean glaciers. Similar examples are also known from North America, where floods scoured out the Channeled Scablands of Washington, the first of such features to be studied, and from the Altai Mountains in central Asia. Flow rates of up to 10,000,000 cubic meters per second (about a cubic kilometer in 100 seconds) have been estimated for these floods (Baker, Benito, and Rudoy 1993). Such flooding is also indicated by bands of mud dropped into marine sediments: this has recorded the drainage of the North American proglacial lakes down the Mississippi into the Gulf of Mexico (Brown and Kennett 1998).

At the end of the last glacial, the Black Sea had been isolated from the Mediterranean and the other oceans for some time by the fall in sea level. During this period much of its water had evaporated, the rate of loss by evaporation having been greater than its rate of input from the local rivers. Evaporation dropped its level by about 150 meters. Approximately 7,500 years ago the rising Mediterranean flowed over the isthmus at Istanbul (now the Bosporus) and flooded the Black Sea. The level of the Black Sea is thought to have risen at a rate of up to about 15 cm (6 inches) per day. The story is, not unexpectedly, not as simple as given here. For the more complete version, one should read the fascinating book by William Ryan and Walter Pitman (1998) of Columbia University. Ryan and Pitman suggest that the flooding might have inspired the deluge legends of the ancient Near East, including the biblical one, as well as led to the spread of agriculture in the Near Eastern region. The latter suggestion is particularly contentious, for if people were driven out and inspired to create legends, there should be evidence that settlements were overwhelmed by the rising waters. Recent discoveries in the Black Sea (Kerr 2000a) have confirmed that at least one structure, apparently a wattle-and-daub hut, now lies at a depth of 91 meters. But as we have seen, other intense flooding occurred near the end of the last glacial, both along seacoasts and inland. And severe floods, even if not as catastrophic as some in the Pleistocene, still occur. At least three are recorded in archaeological sites in Mesopotamia (summarized in Oppenheimer 1998), much closer to where the deluge legends are believed to have originated. So there is no lack of candidates for such inspiration, and working out just which one inspired the legends—if indeed there was only a single inspiration—is likely to be a difficult, if fascinating, task.

The "Black Sea Flood" of Ryan and Pitman makes an appealingly dramatic story. Suspicions of "too good to be true" might even come to mind. And so, alas, it may be in the light of recent work by Ali Aksu, at the Memorial University of Newfoundland, and colleagues in Canada, England, the United States, and Turkey (Aksu et al. 2002). To recapitulate, Ryan and Pitman's hypothesis is that starting about 14,000 years ago, the level of the Black Sea fell because evaporation exceeded the rate of influx of water. As the glaciers melted, the level of the other seas rose, and about 7,500 years ago the sea level became high enough for water to flow into the shrunken Black Sea in a catastrophic flood. Thus, there should have been no flow, either way, between the Black Sea and the Sea of Marmora (which lies between it and the Mediterranean) prior to the flooding, and a flow into the Black Sea immediately thereafter. (This flow later reversed, so that in more recent times water flowed out of the Black Sea.) But Aksu and colleagues found a submerged delta on the floor of the Sea of Marmora, indicating that water was already flowing out of the Black Sea 10,000 years ago. They concede that during the terminal part of the Pleistocene, the level of the Black Sea was well below sea level, but suggest that the level of the Black Sea rose more quickly than the general sea level—although not quite catastrophically—so that the Black Sea flooded outward into the Sea of Marmora and the Mediterranean just before 10,000 years ago, rather than the other way around.

Finally, it might be added that Ryan and Pitman were not the first to suggest a connection between the Black Sea and the Noachian deluge. Their 1998 book was foreshadowed by almost thirty years by British evangelical writer Robert E. D. Clark, who in 1972 suggested that sapropel, a fine-grained organic sediment in the Black Sea, was deposited by the biblical flood.

The rebound of the land when freed of ice mentioned earlier resulted in the drying of rivers and lakes, especially the proglacial lakes. Changes in climate followed deglaciation, particularly increased humidity in some places. In North Africa the Sahara, which had been more extensive during the glaciation than now, was replaced by savanna, supporting even plants such as hibiscus. As the interglacial progressed, these regions tended to dry again, as can be seen from the reestablishment of the Sahara and from the dry riverbeds of Arabia. In central Asia this recent drying had some spectacular results. Fantasy stories about recent prehistoric civilizations, such as the Conan stories, sometimes mention a central Asiatic sea (e.g., Munn 1974). So far as I can tell, no geological sources mentioned such a feature, at least east of the Aral Sea. But recent work (1996) by the Turkish geologist Orguz Erol and colleagues shows that the Takla Makan Desert is the bed of a great lake, once as large as the Great Lakes of North America, all taken together. This lake was apparently more than a kilometer deep in some parts, and its old shorelines and deltas can still be seen. How did the fantasy writers find out about it twenty and more years before Erol's work? I am not certain, but that this desert was once a great lake was deduced by the explorer Sven Hedin and archaeologist Sir Aurel Stein before the 1920s. Hedin had found the remains of boats and piers in what is now an arid region of dunes and sun-baked clay. I had not consulted the archaeological literature for this region—difficult to find by the 1970s—but apparently at least some fantasy writers did.

The ecological effects of deglaciation included the opening of empty new habitat for plants as the ground previously covered by the ice or lakes and thus unoccupied was exposed to colonization. This happened repeatedly, not only with each deglaciation but also during glaciations, as tongues of the ice sheets advanced and then retreated again. Thus, there were opportunities for plants whose seeds (or spores) could be widely scattered and could withstand the somewhat barren soil and the exposure to wind and cold. This produced many hybrid forms "ideally suited to the colonization of areas newly opened to plants" (Stebbins 1950, 348)—in other words, many of what we consider to be weeds.

## Effects on Animals and Plants

The existence of refuges obviously affected the flora and fauna of each subsequent interglacial. Species that previously occupied one broad range, such as the common flicker (*Colaptes auratus*) and yellow-rumped warbler (*Dendroica coronata*) of North America, would have been split into two for the duration of the glacial episode and may have evolved differences during this time of separation. These birds still show distinct eastern and western races. Some plants, too, still remain as separate populations in the West and the East, such as the willow (*Salix vestita*). A recent survey (Klicka and Zink 1997) found that disruption of ranges by the Laurentide ice sheets during the Würm did not contribute significantly to the evolution of new species of birds in North America. However, this was just the latest of several Pleistocene glaciations, and there seems to have been a cumulative effect of all of the Late Cenozoic ice sheets (Avise and Walker 1998).

The geographical effects of the sinking and rebounding land surface, the retreating glaciers, and the resulting shifts of drainage patterns allowed fish and other freshwater creatures to develop discontinuous distributions. For example, the disruption of the river and stream systems of eastern Beringia permitted lake whitefish (*Coregonus clupeaformis*) from the Yukon River drainage to invade the Peel River drainage, where it still survives in Margaret Lake. The Peel is a tributary to the Mackenzie, and the whitefish does not occur anywhere else in the Mackenzie River drainage, since it entered the Peel drainage region when the Mackenzie valley was still filled by the ice sheet. The remainder of the Mackenzie aquatic fauna was established by creatures migrating northward from the Mississippi drainage. (Further details of these kinds of effects in North America may be found in Pielou 1991).

Finally, there was the obvious effect on animal life of developing adaptations to the extreme cold. One of these was the development of the woolly coats of insulation, but at least equally important was the evolution of large body size. Mammoths were usually larger than modern elephants, although the largest African elephants are as big as the largest woolly mammoths (*Mammuthus primigenius*—not the largest of mammoths). And mammoths were not alone in being "mammoth." For example, the Pleistocene Arctic (steppe) bison (*Bison priscus*) were larger than modern plains bison (*Bison bison*) (Guthrie 1990). We don't know whether plants developed such climatic adaptations, but some trees—for example, balsam fir (*Abies balsamea*), trembling aspen (*Populus tremuloides*), and paper birch (*Betula papyrifera*)—can survive temperatures as low as any that now occur.

## Ongoing Processes During the Ice Ages

Many of the events and processes that created the modern world were related to the episodic glaciations—but not all. Mountains continued to rise, volcanoes to erupt, and continents to creep slowly about. All of these processes that had been in train since early in the history of the planet continued, and inevitably some of these also had major effects during the Pleistocene. Preeminent among them were the effects of continental drift.

South America had long been an island continent and hence had evolved a unique fauna. During the Tertiary it drifted largely toward the west, but during the Pliocene, about 4.5 million years ago, a chain of islands that had arisen in the Pacific coalesced and became an isthmus joining South to North America. When this occurred, the faunae mixed; and apparently as a result, many of the South American animals became extinct, while others emigrated and thrived in the north. Even after considerable study, the reasons for the survival of some and extinction of others are still unclear. Most native South American carnivores—which were birds and marsupials—apparently did not survive the arrival of placental carnivores. But among the birds, the giant flightless *Titanis* seems to have survived in North America for about two million years in the face of competition from placental carnivores. Seemingly unlikely forms such as glyptodonts (giant armadillo-like beasts) and ground sloths died out, but in the case of one of the ground sloths, only after successfully migrating as far north as Alaska.

*Orogeny,* the building of mountains, continued as well. This is, of course, not separate from continental drift. The Andes are believed to have elevated substantially, at least in places, during the Pleistocene. The Yulongxue Shan of southwestern China (*shan* is "mountain[s]" in Chinese) show evidence of only the Würm glaciation and are thus thought to have been elevated since the Riss, a conclusion that is perhaps not entirely convincing because the latest glaciation may have obliterated all traces of previous ones. Similar uplifts are thought to have taken place in New Guinea.

And, as always, there was volcanism. The Deep Sea Drilling Project of the early 1970s examined cores of sediment from sea floors around the world. These documented a substantial increase in explosive volcanism during the Pleistocene (Kennett and Thunell 1975). The amount is thought to have been adequate to trigger the development of extensive glaciers, although whether it did so in fact is unclear since continental ice sheets seem to have started about 2.4 million years ago, just at the beginning of the increase in volcanism. But is this really unrelated to climatic changes? With glaciations the sea level falls and the sea floor may respond to the decreased weight of the (now smaller) ocean. Part of this response may be an increase in volcanism. This certainly seems feasible, but whether it happens (and how often) is another matter.

The terrestrial magnetic field has seemingly altered. In fact, some work indicates that the field was substantially (perhaps 25 percent) weaker during the height of the glaciation than it is now. Whether this was directly related to the glaciation is not known, especially since the mechanism producing the earth's magnetic field is itself not completely understood, although it is thought to depend on events at the earth's core, not at the crust. And there is evidence that—then as now—there was from time to time exceptional (often exceptionally bad) weather. One example is an exceptionally intense monsoon that was recorded in the sediments of the Mediterranean Sea and Indian Ocean about 537,000 years ago (Rossignol-Strick and Planchais 1998).

## Animals of the Ice Ages and Their Fossils

Of the world's fauna during the Pleistocene, the African mammals would have seemed familiar, although there were also some unusual ones, like deinotheres, large elephantine beasts (related to elephants) with downward-curved tusks in their lower jaws. But Africa, by and large, has been a refuge for animals of Pleistocene type. We would have seen these kinds of animals elsewhere as well, from the British Isles through continental Europe to tropical Asia. A good introduction to these Pleistocene mammals may be found in Bjorn Kurten's 1971 book, *The Age of Mammals;* here we can only sketch out their variety. As mentioned before, elephants and hippos lived in Europe during the interglacials, and woolly rhinos during the glacials. There were also animals unlike modern ones, such as the elasmotheres, great rhino-like animals with a huge single horn projecting from the forehead (worked into a poem by the Dane, Johannes V. Jensen). There were many large beasts, larger than their current relatives, but some, like the "donkey elephants" of Malta, had evolved to become smaller. The faunae of Africa, Eurasia, and North America had been in long but periodic contact. South America had evolved a fauna of unusual-appearing beasts, herbivorous mammals superficially similar to those of other lands. These included toxodonts (similar to hippos), astrapotheres (similar to elephants, but smaller) and notoungulates (similar to horses). The carnivores were marsupials, some superficially similar to those found elsewhere, and giant flightless birds. In Australia, the herbivorous mammals were marsupials—diprotodontans, kangaroos, and palorchestids (described in the next chapter)—and the major predators seem to have been reptiles. In New Zealand, which had been considerably more isolated than Australia, the large herbivores were all flightless birds, and the predators were flying birds.

These animals bring to mind the famous tar pits at Rancho La Brea (now in Hancock Park) in southern California. We have all seen the restorations of now-extinct beasts trapped in the soft viscous tar, and small mammals and birds are still trapped by the tar seeps. It is an image of a perilous primeval past in which simply walking on the ground in the wrong place was dangerous for the unwary. In their 1973 study of these deposits, Geoff Woodard and Leslie Marcus found them to date back to 35,000 years ago. Many bones are weathered, worn, and abraded, which must have happened in flowing water and could not have occurred while the bones were embedded in the asphalt. The bones are found, not in asphalt pockets, but in asphalt-impregnated clays and sands. Fossil mollusks suggest that the asphalt was often covered by the shallow waters of lakes or rivers and thus not obviously accessible unless the animals were wading. Furthermore, in the lowered temperatures of the Würm, the asphalt must have been more solid than at today's temperatures, just as asphalt roads may soften in the summer but remain solid through the winter. Woodard and Marcus, although not denying the episodic existence of traps of asphalt-soaked quicksand, show that many of the bones accumulated in rivers, from carcasses washed in or animals trapped in soft sands and muds. The asphalt was later forced in from the deeper petroleum-bearing rocks. Bones have been found in "essentially asphalt-free sediment" (Woodard and Marcus 1973, 66) on the edge of Hancock Park that clearly accumulated in a stream. Thus, the deposits seem to have at least two origins: traps of asphaltic quicksand (the "traditional" view) or thin layers of liquid asphalt, but also the preferential preservation of bones that accumulated in stream channels simply because the sediments were later soaked with asphalt. There are other asphalt seeps in California (McKittrick, Carpinteria, and others), as well as in Trinidad, the Andes, and the Caucasus, but these have not been the subject of as much study as those of Los Angeles.

The "muck beds" of Alaska (and Siberia), where more-or-less complete carcasses have been found buried in mud and silt, are superficially similar. But these occurrences seem to have resulted from the carcasses of dead animals being covered by "creeping" soil and then preserved in the permafrost (Guthrie 1990). Some carcasses preserved here, but more especially those preserved in similar beds in Siberia, have been said to have been "quick-frozen." In other words, to have died, not from obvious causes, but from extreme cold, much colder than could be explained scientifically. This sensational conclusion was based on finds of fresh-appearing vegetation in the mouths or stomachs of carcasses and the apparent freshness of the meat. During his visit to Australia, Andrei Sher, a Russian paleontologist who specializes in the Siberian Pleistocene, described a first-hand encounter with such a carcass. The meat did appear fresh and red when it was first uncovered, he said, but it smelled far from fresh even then, and it rapidly turned an unappetizing brown. These seem to be carcasses of animals that became trapped in mud or fissures in the ground. The fresh-looking vegetation still in the mouth probably resulted, not from "snap freezing," but from a trapped and hungry animal trying to feed to the last on whatever it could reach. Unfortunate incidents to be sure, but nothing paranormal.

Many of the Pleistocene animals of Alaska and of the Yukon represent the eastern extremity of a fauna extending west across all of Asia to the European ice sheet. Saiga (*Saiga tatarica*), antelope now found only in central Asia, ranged to the Yukon. Northern Siberia, Beringia, Alaska, and the Yukon supported a fauna of large mammals: bison, mammoths, horses, and (at least in Asia) rhinos. The paradox of Beringia is that there was a diversity of animals, many of them large, but the plant fossils do not indicate sufficient numbers or variety of plants to feed them. Neither modern boreal forest nor tundra supports faunae of large mammals—so what kind of habitat supported these animals during the Pleistocene?

## Paleoecology During the Pleistocene

The research of Sher (1996) in eastern Siberia has posed several unexpected questions. First, he could find no clear evidence of changes in the fauna or flora during the interglacials. This is quite unlike what happened in Europe and North America, where there were substantial changes. Second, the pollen data suggest that the Pleistocene plant communities, both glacial and interglacial, were distinct from those now living in that region. Third, there is no clear evidence of substantial marine transgressions, invasions of the sea, that should have resulted from melting glaciers during the interglacials.

Sher and others have argued that the Arctic Ocean did not rise substantially at these times and so remained a region of permanent ice cover, summer and winter, glacial and interglacial. The restriction of the Gulf Stream to the southern part of the North Atlantic, and the absence of its flow northward into the Arctic Ocean (as the North Atlantic Current) would presumably have contributed to the permanence of this ice cover. The permanent ice cover of the Arctic Ocean and the blocking of the westerlies resulted in drier, more continental climatic conditions than now exist for eastern Siberia—a region which, together with Beringia and Alaska, he terms Beringida. There was too little humidity for the development of an extensive continental ice sheet, although there would have been considerable permafrost and some highland glaciers. Beringida would have been frigidly cold during the Arctic winter-night but may have heated up to almost tropical temperatures during the 24-hour-long sunlight of summer. The height of the European ice sheets would have blocked westerly winds from reaching Siberia and resulted in a permanent high-pressure air mass (anticyclone) in this region. This air mass is thought to have produced higher air pressures than are now found on earth, and that, in turn, would have reduced cloud cover in the summer (Velichko 1984)—hence the "almost tropical" temperatures of this season and, presumably, steep temperature gradients onto the neighboring ice sheets. This would have been an almost Martian environment, although on Mars the temperature range is daily, not seasonal, and the air pressure is much lower. This plain was inhabited by mammoth, woolly rhinos, five or six kinds of horses and asses, musk ox, reindeer, and Saiga, with several kinds of deer in the southern river valleys (Ukraintseva 1993). Beringida was a habitat unlike any now existing—a "non-analog" habitat. Sher, among others, believes that Beringida was an immense grassy steppe, the mammoth steppe, with some sparse shrub or tree cover that fluctuated in extent.

This non-analog ecosystem included species that still exist today but that now live in different habitats. This has led some scientists—in view of the absence of reliable dates—to suggest that the simplest explanation is that there *was* no non-analog ecosystem. Instead, they have suggested that the disparate forms involved actually lived at different times in ecosystems essentially the same as those they inhabit today. Although there was already much evidence in favor of Sher's view, a recently published large group of dates from North America and Asia (Stafford et al. 1999) shows that disparate forms of mammals did live in the same places at the same times. Thus, Sher's view seems likely to be correct.

An implication of Sher's argument is that our current Holocene conditions are fundamentally different from those of the past million or so years. A corollary is that the extinction of the Beringidan mammalian fauna was due to the "collapse" of the steppe plant community following climatic change about 12 millennia ago.

## How the Ice Ages Happened

*The Astronomical Mechanism*

Much of the reason for the foreign feel of the Pleistocene is the different climatic conditions. A Serbian scientist, Milutin Milankovitch (1879–1958), worked out how these climatic changes came about. He reasoned that the rhythm of the glacials and interglacials was not something intrinsic to the earth at all but—like most of the earth's light and heat—was imposed by the sun. Thus, he made detailed calculations of the earth's orbit. The orbits of the planets are elliptical, but orbits can be changed. With spacecraft this is usually done by the use of rocket motors, but planets' orbits also change, albeit only slightly, from interactions with the gravitation of the sun and other planets (in Earth's case, largely Venus). So the terrestrial orbit is not an immutable ellipse, but it changes, stretching by as much as about 18.27 million km. Astronomers know this as a change in *eccentricity*, which is an index of the amount by which an orbit deviates from circular. When a planet is further from the sun, it gets less heat and so cools; and when nearer the sun, it warms. If its orbit changes so that the planet comes to move at a greater distance from the sun than it did in the previous version of its orbit (Fig. 1.5), the planet will cool more than it had before and perhaps initiate a new and cooler climatic regime. Milankovitch proposed that this kind of effect was the ultimate cause of the ice ages. Well, part of the cause, as it turned out.

The earth, of course, has other motions in space than just orbiting around the sun. It also spins on its axis. This spin is not an entirely even motion: the axis "wobbles," each "wobble" taking about 25,800 years. This wobble is termed *precession*. When the Northern Hemisphere—having more land area than the Southern—is tilted toward the sun in the winter, the land does not cool as much as when the axis is tilted toward the sun during the summer (Fig. 1.6). In other words, having a cooler summer and warmer winter counts more toward maintaining an interglacial climate than having a hotter summer and colder winter. So if the axis wobbles from having the Northern Hemisphere toward the sun in the winter to having it toward the sun in the summer, the land areas, the whole hemisphere—and hence the whole earth—will cool.

Finally, the tilt of the axis changes, from 21.5° to 24.5° (Fig. 1.7). The seasons are more extreme when the axis is more strongly inclined, and the greatest inclination happens every 41,000 years.

What Milankovitch supposed was that (1) when the earth's orbit shifted to take it further from the sun, and (2) the axial tilt brought the Northern Hemisphere closer to the sun in the (northern) summer, at the same time that (3) the axis was more strongly tilted, the climate could chill enough to bring on an ice age. These are three independent motions, and so do not necessarily occur together; but when they do, it should get cold. Calculating when these three conditions occur together took Milankovitch years of computing by hand. Each state of the orbit and inclination of the axis will have a different degree of effect on the climate. Simplifying each of these factors to two conditions, one that heats the earth and the other that allows it to cool, we have three separate influences, each of different strength, that can occur in eight different combinations. So they can provide eight different degrees of heating or cooling. The weather, being noted for its variability, would tend to blur these into a range of responses. This being the case, how can one work out whether or not this is a plausible cause?

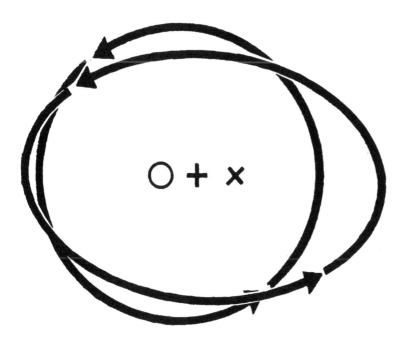

Fig. 1.5. Changes in the eccentricity of the earth's orbit. Eccentricity is a measure of how much the orbit deviates from circular. Thus, the more nearly circular orbit has less eccentricity than the more elliptical orbit. All orbits (in the Solar System, anyway) are elliptical, and so have two foci. On the diagram the focus occupied by the sun is marked by a circle, the second of the less eccentric orbit by the + and the second of the more eccentric orbit by the x. (From Kurtén 1972.)

Milankovitch did this by plotting all these together and calculating the resultant temperatures of the earth's surface. He found that the coolest periods worked out to roughly match the timing of the ice ages. Milankovitch's work was not taken very seriously for two main reasons. First, no one wished to go to the trouble of repeating his calculations by hand to test them, an unbelievably long and tedious chore. And second, there were not enough good data on paleotemperatures and climatic history with which to test them anyway. Now that we have computers to do the calculations and much better data, we find that Milankovitch seems to have been right—his calculations match the data well. So this is generally taken as the ultimate cause of the ice ages.

Notice that Milankovitch predicted ice ages throughout the earth's history. At the time he worked, most attention was given the Pleistocene glaciations, although geologists—especially in the Southern Hemisphere—were aware of the Permo-Carboniferous glaciations (some 250–350 million years ago). As noted previously, data gathered over the past forty years confirm that glacial episodes occurred throughout terrestrial history. However, if Milankovitch's mechanism were the only relevant factor, there would have been a more obvious and unbroken sequence of glacial advances and retreats. Other factors must have played some role. Milankovitch's mechanism seems to be the underlying cause, but it is mediated by other effects that act together with it to produce ice ages.

### Rates of Change

There is always something new to be learned. Milankovitch said nothing about how fast the climate changed from an interglacial to a

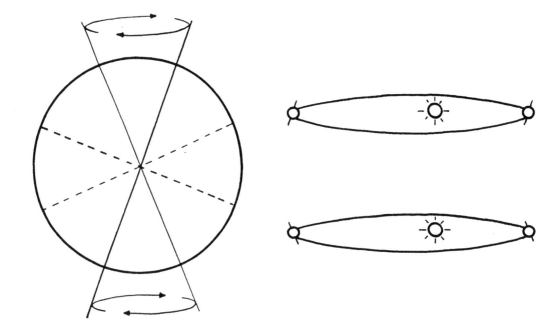

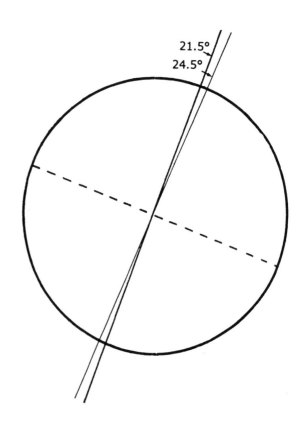

Fig. 1.6. (above) The axis around which the earth spins, itself moves. During this motion, precession, the axis sweeps out a double cone, indicated at the left. The circle represents the earth; the continuous lines, two opposite positions of the axis; and the dashed lines, the corresponding positions of the equator. To the right, we see the earth at its closest approach to the sun (perihelion) and its farthest departure (aphelion). Precession results (eventually) in changes in the relationship of the axis to the sun. When the Northern Hemisphere—having more land area than the Southern—is tilted toward the sun in the winter, the land does not cool as much as when the axis is tilted toward the sun during the southern summer. (Adapted from Kurtén 1972, and Chorlton 1983.)

Fig. 1.7. (left) In addition to precessing, the tilt of the axis changes, from 21.5° (heavy line) to 24.5° (light line). The equator, at the lesser tilt, is indicated by the dashed line.

glacial one or, for that matter, just how this happened. It was generally felt that climatic change would have been reasonably slow, if only because there is a lot of air in the atmosphere, as well as a lot of water in the ocean, to heat or cool. When the glacials ended, all the water frozen into the glacial ice caps had to melt, and this too must have taken time. Thus, the findings of Genevieve Woillard at the Catholic University of Louvain in Belgium were astonishing. In 1979, she announced that the transition from interglacial to glacial conditions in northeastern France happened relatively quickly. Her study of pollen from the Eemian transition indicated that temperate forest—rather like modern French forests—was replaced by taiga (a conifer forest) in 150 ± 75 years (i.e., in 75–225 years). This is within three (admittedly long) human lifetimes.

Perhaps this should not have been as much of a surprise as it was, because almost twenty years earlier Wallace Broecker and his colleagues Maurice Ewing and Bruce Heezen, all well-known oceanographers from Columbia, had suggested (1960) that the Pleistocene climate underwent abrupt changes. Ten years after Woillard's paper, W. Dansgaard at the University of Copenhagen (Denmark), J. White at the University of Colorado (USA), and S. Johnsen at the University of Iceland completed their study of evidence from the Greenland ice sheet for climatic change immediately after the last glaciation. In the meantime, further evidence from cores drilled from the South China Sea supported the notion of abrupt climatic shifts. So when Dansgaard and his colleagues found that shifts from cooler to warmer climate occurred in less than twenty years, no one was surprised. They found that the Younger Dryas (the last cold period at the end of the Pleistocene) ameliorated in about two decades, and southern Greenland warmed 7°C in approximately fifty years. Since then, more evidence has come to light, and shifts of climate during a human lifetime, though still catastrophic, do not seem outlandish.

Ice cores from the Greenland ice sheet—near the center so that they were not distorted by the results of the flow of the ice to the sea—were examined to find out how the climate, at least of Greenland, had changed over the past 150 millennia. Temperature could be ascertained from stable isotope ratios, like $O^{16}$ to $O^{18}$ (which were, in fact, used). Evidence known then and collected since, from the cores of pond, lake, and sea-floor sediments from many sites in the Northern Hemisphere, indicated that what was seen in Greenland was reasonably representative of the whole hemisphere—taking into account changes in temperature with distance from the pole, of course. The results out of Greenland clearly showed that the glacial climate was not as stable as nearly everyone had assumed.

### The Role of Atmospheric Composition

Climatic change seems to be largely a matter of heating or cooling air, but just how does this happen? Is it simply by adding or losing heat, like boiling water for tea in a microwave? Or are there complications? The composition of the air does play a role, a complication known as the *greenhouse effect*. Greenhouses maintain high internal temperatures by passing sunlight through their glass roofs and walls. Some of the incoming radiation is absorbed by the plants, soil, and such and heats these, but some is re-radiated as heat (infrared) radiation. Glass, however, is less transparent to this back radiation than it is to incoming solar radiation. Thus, the back radiation is trapped in the greenhouse, heating it. Greenhouse gasses act in a similar manner. Their molecules are "transparent" to incoming solar radiation but absorb the re-radiated back radiation, which they then radiate again, heating both the air and the ground surface.

The greenhouse effect raises the average temperature of the earth's surface from the −18°C that would prevail in the absence of greenhouse gasses to a more tolerable 15°C. Greenhouse gasses include carbon dioxide, water vapor (the most abundant), and methane.

The amount of water vapor, linked to the amount of water in the oceans and glaciers, seems not to change much over periods shorter than hundreds or thousands of millions of years. So it plays little role in the creation of the much shorter ice ages. Carbon dioxide is a different matter. The removal of carbon dioxide from the air should act to cool the earth and, so far as we can tell, it does. Over most of the earth's history, its surface temperature has remained remarkably constant compared to those of Venus and Mars. Unlike Mars and Venus, the earth has a mechanism for regulating the amount of carbon dioxide in the air, the carbon dioxide cycle, sometimes called the *chemostat*. Carbon dioxide is released into the air by volcanic eruptions and by the solution of submarine carbonate deposits. It is also released by the respiration of microorganisms, animals, and plants and absorbed by photosynthesis. It is absorbed, too, by the deposition of carbonate sediments, limestone, and the like and the erosion of rocks. In erosion, carbon dioxide combines with silicate minerals that include calcium to produce carbonate minerals and silica. The deposition and solution of carbonates in the oceans is sensitive to the amount of carbon dioxide in their waters; and that amount, in turn, is sensitive to the amount in the air. Thus, if the amount in the air increases, the deposition of carbonates increases, but if the concentration decreases, carbonates dissolve and release carbon dioxide. This keeps the concentration of carbon dioxide in the air stable—within bounds. If there is a massive drain, the carbon dioxide concentration will fall, and this will decrease global temperatures. This kind of effect may well mediate between the variations in the solar radiation caused by the orbital vagaries studied by Milankovitch and the global temperature, and hence the climate as a whole. But there is more to the story.

Methane is a much more potent greenhouse agent than carbon dioxide. Over the past several years, partly as the result of offshore drilling for oil, it has become clear that large methane deposits occur in the sea floor. It may seem odd to write of deposits of a gas, but the methane exists as methane hydrate, a form of methane combined with water molecules, giving a solid, not a gas. At temperatures below about 5°C and depths below about 300 meters, this exists as a white, icy substance. It has been estimated that much more methane exists in this form than as a gas in the atmosphere.

If the sea floor is disturbed, as by an earthquake or even by warmer-than-usual currents, the methane may be released and bubble to the surface. This happened several times in drilling for oil in the Gulf of Mexico and the North Sea. It has even been suggested that eruptions of methane from the Atlantic floor off Florida, where methane hydrate is reportedly common, may account for mysterious disappearances of vessels in the "Bermuda triangle." A ship trying to sail a sea that is abubble with methane loses considerable buoyancy as the sea itself becomes less dense from the bubbles, and it founders.

The floor of the Norwegian Sea west of Norway bears a field of craters that are up to 3.3 km in diameter. Jürgen Meinert in Germany found that they were formed by the release of methane from sea-floor deposits about 8,000 years ago, as the latest continental ice sheets were melting. The amount of methane released must have affected the climate and possibly accelerated the rate of melting. Recent work off the coast of southern California indicates that methane was released during several interstadials, perhaps because of the warming of the overlying sea water (Kennett et al. 2000). These injections of methane into the atmosphere may have even been responsible for the interstadials. A large submarine landslide in the western Mediterranean 22,000 years ago may also have released methane (Rothwell, Thomson, and Kaehler 1998). Thus, it seems likely that methane, as well as carbon dioxide, has affected the global climate.

## Geographical "Triggers"

Given the variability in the amount of heat received from the sun by the earth as the ultimate cause of glaciations, we still have to explain just how they are started. Apparently, not every time that solar heat has fallen to a minimum has been marked by glaciations; there seem to have been none during the Mesozoic (the "age of reptiles"), for example. The position of the continents may be relevant. During much of the Mesozoic, both north and south polar regions were (at least partially) occupied by seas through which warm currents may have flowed and which could not accumulate snow as easily as land does.

So we see that the climate does not respond directly to the input of solar radiation. The atmosphere, oceans, and land masses all act as reservoirs of heat and influence how the climate responds. As we have seen, just how the atmosphere responds depends on its composition. The orbital variations occur relatively gradually and so, we might think, should their effects, such as causing an ice age. Nonetheless, the climatic changes creating an ice age can occur rather rapidly even if the accumulation of snows into ice sheets takes longer. And the initiation of ice ages themselves may well involve geographic factors, such the arrangement of the continents.

Antarctica developed an ice sheet early in the Late Eocene (by 38 million years ago) that may have lasted—with a few short interruptions—to the present day, regardless of the varying input of solar radiation. This ice sheet probably formed as the result of the splitting of Antarctica and Australia, the last vestiges of the southern supercontinent, Gondwanaland. When this happened, the southern ocean encircled Antarctica, and cold currents swept unimpeded around that continent, thus maintaining around it a reservoir of cold water. Previously, the currents were deflected north into temperate or tropical regions, cooling these regions but warming the Antarctic when the waters returned south as warm currents.

There are other ways in which the disposition of land can affect climate. Major climatic change during the Tertiary has been attributed to the uplift of the Tibetan plateau. This plateau is the largest on the earth, occupying almost half of one percent (0.4% to be exact) of the earth's surface, with an average altitude of 5 km. This land was elevated by the collision of India with the Asian mainland that started sometime around 65 million years ago and still continues today, with the Assam earthquake of 1897 elevating the land by 11 meters (Bilham and England 2001).

The process by which this uplift altered the climate is perhaps not what one would immediately suppose (unless one happened to be a geochemist). There seems to have been some alteration in the circulation of the atmosphere over Eurasia and India, but this played only a small role in the ensuing climatic change. The chief effect was due to increased erosion. We often think of erosion as primarily a physical process, the breaking and washing away of rock by flowing water ("Little drops of water, little grains of sand, run away together, wear away the land"). Or perhaps the basically similar sandblasting effect of wind erosion comes to mind (more difficult to express as doggerel). But erosion is also a chemical process. Certain atmospheric gasses are absorbed, and hence removed from the air, in the process. For some gasses, like nitrogen, this may not lead to any obvious effect on the climate, but it is carbon dioxide, not nitrogen, that is involved. A massive drain of atmospheric carbon dioxide is thought to have occurred as the result of erosion in the rising Tibetan region. Enough of the gas was absorbed by the rising lands that the total atmospheric concentration fell, regardless of the carbon dioxide cycle releasing it from carbonates.

As already mentioned, there were also changes in the circulation of the air. Computer simulations indicate that the Indian monsoon is largely due to the presence of the Himalayas (Paterson 1993). Other such effects include creating a drier climate in Mediterranean region and in central Asia (and hence creating regions like the Takla Makan Desert and the Gobi), a cooler climate in Europe, and a wetter one in southeastern Asia.

The most recent suggestion is that Australia was partly responsible for the ice ages—not directly, but by the northward movement of the New Guinean region. As New Guinea approached Indonesia, the intervening channel became shallower and narrower. This reduced channel diminished the amount of warm water flowing west from the tropical Pacific into the Indian Ocean. The main effect, however, was that the northward movement of New Guinea blocked the flow of the south equatorial current. This, in turn, reversed the flow of water to direct it to the east as the equatorial countercurrent. The heat distribution through the Pacific was altered to increase the amount of cooler air flowing from the ocean into Canada, where the precipitation accumulated into ice.

Other possible triggers for glaciations include volcanism spreading dust through the atmosphere and so reflecting sunlight and reducing the temperature, and variation in the heat put out by the sun (in contrast to variation just in that received by the earth). Given the suite of different frequencies of the orbital and rotational effects, the climatic results can show a superficially random-seeming pattern. However, in spite of this they, like other celestial motions, are reliably predictable if one considers sufficiently long periods of time.

## Mechanics of Climatic Change

The detailed mechanics of how the climate changes during the coming of the ice sheets is still unclear, although it is rapidly becoming better understood. The variations in sunlight affect different latitudes differently, as we might expect from the different angles at which sunlight falls on the earth's surface at different latitudes. Climatic changes seem to start in the Southern Hemisphere (White and Steig 1998) in the heating and cooling of the southern ocean. The heat is thought to be transferred between hemispheres by oceanic currents such as the so-called "conveyer belt," a current linking the Northern and Southern Hemispheres and hence their climates. This current arises from cold water of the North Atlantic sinking off Greenland and flowing south along the eastern coasts of the Americas, eventually turning eastward to cross the South Atlantic to the sea south of the Cape of Good Hope. From there it flows east through the southern ocean, and east of New Zealand it turns again to flow north, eventually into the North Pacific. As it flows, the water slowly warms, so by the time it reaches the North Pacific it rises to the surface and then flows back through much the same course to arrive in the North Atlantic. The initial sinking, incidentally, is not caused so much by the coldness of the current as by its increased salinity, resulting from evaporation as it flows northward (Fig. 1.8).

The most recent research (Shackleton 2000) found a good correlation between the amount of sunlight received on earth (as predicted by Milankovitch) and the atmospheric concentration of carbon dioxide. So good, in fact, that some kind of direct influence of one on the other has been suggested. What the link is, admittedly, is unknown, although the warmth of the oceans, and hence of currents like the "conveyer belt," is a reasonable possibility. Regardless, it seems that (1) variations in the earth's orbit affect (2) the amount of sunlight reaching the surface, which (presumably by heating) affects (3) the concentration of greenhouse gasses in the air, that in turn affects (4) the amount of ice accumulating on land and sea. The first three items act pretty much in synchrony, but the ice accumulates (and disappears) more slowly.

## Climatic Instability

As already mentioned, the glacial climate (and perhaps the climate in general) seems somewhat unstable. There have been abrupt but transient shifts during which the climate of the Northern Hemisphere warmed within decades, then cooled again more slowly, sometimes for as long as a few centuries. These periods seem to have lasted for from about seventy years to approximately a millennium (Stocker 1998). Twenty-four such shifts, or Dansgaard/Oeschger oscillations, occurred during the Würm. Until 25,000 years ago, these occurred approximately—but only approximately—every two millennia. These warmings seem to be related to great abundances, or "flotillas" as they have been called, of icebergs in the North Atlantic, termed "Heinrich events." Hartmut Heinrich in Hamburg found that the sediments of the North Atlantic show layers of rocky debris brought by icebergs from North America. These layers apparently mark times in which the Laurentide ice sheet was breaking up along its seaside edges to calf into fleets of icebergs.

However, Heinrich events were not limited to times when the ice sheet was melting. Current research suggests that changes in the circulation of the North Atlantic, and probably also the Pacific, are involved in forming and maintaining the ice sheets (Cane 1998). These changes may involve altered circulation in which warm currents, such as the warm part of the "conveyer belt," move more or less far north (in the Atlantic) or warm surface waters more or less far east (in the Pacific). The climatic changes were not restricted to the North Atlantic region but correlate with changes in the vegetation in Florida, glacial advances in the Chilean Andes, temperature fluctuations in the Antarctic, and increased winds in the South Pacific near South America, suggesting that the entire global surface is affected.

The climatic oscillations revealed another complication. Although the Milankovitch mechanism involves a number of rhythms (Table 2), cores from both glaciers and the sea floor showed still further cycles. Some of these, at least, seem to be overtones (that is, cycles whose periods are one-half, one-third, one-quarter, etc. of the basic cycle) of the fundamental Milankovitch cycles, particularly that of precession. One cycle, with a period of about 11,000 years, may result from the equatorial lands being warmed by the warm phases of the precessional cycles of both hemispheres, thus giving a cycle with twice the frequency.

The earth's climate is governed by the interaction of external and internal, astronomical and tectonic factors. During ice ages, it seems to be a two-phase system that shifts from one phase (interglacial) to the other (glacial), depending on the amount of solar radiation received. During the different phases, the temperatures of the sea surface change, not so much in range as in the geographical extent of warm versus cool surface water. These changes affect the climate of the overlying atmosphere and the concentration of greenhouse gasses, and that, in turn, affects the climate. Pack ice in the southern and Arctic oceans accumulates and reflects solar heat, thus cooling the overlying air and the adjacent lands. Presumably as the result of this cooling, possibly augmented by increased cloudiness, ice sheets up to 3 km thick spread over those continents near the poles. Just why is still unclear, but immediately before the spread of the ice sheets, there was a marked increase in frequency of climatic cycles (Willis et al. 1999). As the climate was cooling, this may have allowed some ice that accumulated in the cold

Fig. 1.8. The "conveyer belt," the currents that convey and distribute heat throughout the world's oceans. Black arrows show the cold, dense, salty deep sea currents; and white arrows, the warmer, less saline surface currents that return water to the North Atlantic.

periods to remain unmelted through the now shorter warm spells, thus building into glaciers. Most of North America (including Greenland) and Europe were covered, and all of Antarctica. Antarctica remains like this today, and the present ice sheet there may have continuously existed for more than 3 million years.

## Climate as a Two-Phase System

The climate of the earth in general, and not just during ice ages, is believed by some geologists to alternate between two states, evocatively (but not very informatively) termed "greenhouse" and "icehouse." The glacial and interglacial climatic regimes of ice ages would thus be phases of the icehouse climate. Greenhouse and icehouse states are best known for the Phanerozoic and Late Precambrian, although presumably they existed before this period. Icehouse climate existed from Late Precambrian to Late Cambrian, Late Devonian to Late Triassic, and from the middle of the Tertiary to the present. The intervening periods enjoyed a greenhouse climate. Each state was maintained by the concentration of greenhouse gasses, particularly carbon dioxide: high (thought to have been about twice current values) for the greenhouse state, and low (at current value) for the icehouse. During the icehouse state, as now, there were substantial variations in temperature from equator to pole, and hence icecaps at the poles; relatively low sea levels, hence widespread continental climates; and cooler oceans with more active circulation and greater levels of dissolved oxygen. The green-

Table 2
Frequencies of various components of the earth's motion in space

| Frequency (once/x years) | Aspect of earth's motion |
|---|---|
| 1,900,000 | orbital shape and orientation (minor component) |
| 70,000 | orbital shape and orientation (major component) |
| 50,000 | orbital shape and orientation (second minor component) |
| 41,000 | obliquity (axial "tilt") |
| 25,800 | precession (axial "wobble") |

house state, known only by inference, is characterized by higher global temperatures, and hence warmer, ice-free poles; and relatively high sea levels (as less water is "locked up" in ice sheets on land), hence milder, more maritime climates. There were more extensive epeiric seas (those flooding over continents), and warmer oceans with less circulation and lower levels of oxygen.

The times of these episodes are by no means agreed. While some would put the beginning the current icehouse state at about 20 million years ago, others put it as recently as 3 million.

## Heritage from the Pleistocene

We tend to the think of the past as being essentially like today, and of course it was. The trick, however, is in the word "essentially." The actors, the laws of physics and the processes of geology, have not changed; but the theater, the conditions on the earth, has. We easily mistake the one for the other. The guideline of *uniformitarianism*, specifically of *actualism*, often used in geology (and evolutionary biology), is that the processes now occurring are those that have always occurred, and that processes that took place in the past can still take place (and so be studied) now. This was opposed to the notion of *catastrophism*. Some scientists now espouse a kind of scientific catastrophism that recognizes catastrophic events of the kind that happened during the Pleistocene. This, however, is not the original notion of catastrophism, which invoked catastrophes that suspended the laws of nature and involved divine (as in the Noachian flood), or at least supernatural, intervention. We have intellectually moved so far from invoking supernatural influences that, unless they are also "scientific creationists," those who invoke catastrophes clearly accept actualism.

The conditions of the Pleistocene, particularly during the glacial times, were different—in ways that we may not find obvious—from those of the present. This chapter has discussed some of these differences, and they are further discussed in the many books on the Pleistocene, of which Goudie (1977) is a good starting point. We should not forget that there may be yet other differences that we do not recognize. Much of the feeling of difference from Pleistocene times is related to the substantially different environmental conditions. The extensive ice sheets are beyond the experience and perhaps even the imagination of many of us. But there were not only these impressive long-term differences but also a series of ephemeral but catastrophic events that were probably repeated with each retreat of the ice sheets. These included the rapid and extensive flooding directly due to the melting of the ice like that which formed the Channeled Scablands, and also the flooding of the seacoasts due to the general rise in sea level. There were also catastrophes of other kinds: the drying of the vast proglacial lakes when they drained (catastrophic to their inhabitants at least, if not necessarily to us), and the droughts associated in some regions with the drier climates of the glacials. The history of the Pleistocene reminds us that the physical environment is not simply the stage for our dramas but is itself a part of the cast whose role we cannot take for granted.

The real point of this chapter, however, is to show just how different the Pleistocene past was from the present. Those who live in Europe or North America can imagine from this description just how much their lands have changed. So can those who have visited these continents or even seen them on television travelogues. In Australia, where we will try to understand the paleobiology of *Megalania*, much less is known of the Pleistocene conditions. If the conditions were so different in the northern continents, they may well have been equally different in Australia. But we don't know the details, and this uncertainty about its environment affects some of our conclusions about this great lizard.

*The past is a foreign country . . .*

—Hartley, 1967

# 2. The Pleistocene in Australia

## Introduction: The Southern Hemisphere

The events and conditions of the Pleistocene of the Southern Hemisphere are less well understood than those of the Northern Hemisphere. Climatic fluctuations occurred more or less simultaneously in both hemispheres, with the southern probably leading the northern by a millennium or so (White and Steig 1998). The Southern Hemisphere has less land than the Northern, and the continents are located mostly in the subtropical to tropical climatic zones—or, in the case of Antarctica, right at the pole. Temperate land masses like those that supported the vast ice sheets in the Northern Hemisphere are largely (though not entirely) missing. Thus, glaciation was restricted to the mountains in Australia and South America (as well as New Zealand and New Guinea). In Antarctica glaciation was almost continuous: the advance and retreat of continental ice sheets seen elsewhere did not seem to occur there. The Antarctic ice sheet did retreat or disappear at least once, perhaps even episodically, but that happened well before the Pleistocene.

## History of Land-Dwelling Vertebrates in Australia

Australia has been slowly creeping northward since its final break with Antarctica during the Eocene, as if determined for a vacation in the Tropics. Before the Eocene, about 55 million years ago, the land that would become Australia was situated near the southern polar region for at least 250 million years. A concatenation of events, all more or less closely linked to its geography, led to Australia's having what is probably the most incomplete fossil record of any continent. First, Australia is centrally situated on its tectonic plate (plates are segments of the earth's crust that may carry continents, and whose slow motion causes continental drift) and hence is the flattest of the continents. Mountains are generated near the edges of plates, especially the leading edges, in this case in New Guinea and New Zealand. The result is that whereas

Fig. 2.1. A Cretaceous invertebrate fossil (belemnite) locality west of Tambo, central Queensland. The (absence of) relief shown here is typical of Mesozoic and Cenozoic fossil sites in Australia.

in other lands outcrops of fossil-bearing rock may be measured in tens or hundreds of meters, in Australia a vertebrate paleontologist who locates a meter (about three feet) of outcrop is envied for his amazingly good luck (Fig. 2.1).

Secondly, late in the Cretaceous, Australia passed over a piece of crust sunken (subducted) into the mantle (Gurnis, Mueller, and Moresi 1998), which elevated the entire continent by about 250 meters (about 820 feet) at that time. Consequently, weathering penetrated deeply into Mesozoic sediments. In mid-1998, together with Paul Sereno, I took the opportunity to examine exposures of the Winton Formation (about 95 million years old) in southwestern Queensland. This formation has yielded a few dinosaurian bones in central Queensland and has also produced extensive dinosaurian trackways there. But although the strata we examined showed sedimentary structures where ancient stream channels had been cut and later filled, and yielded pieces of silicified wood from time to time, there were no bones of any kind, not even the smallest of fragments. Such barrenness is characteristic of Australian continental sedimentary rocks older than the Miocene (about 24 million years).

And finally, because of its generally low topography and its location too far north to be affected by Pleistocene glaciations and far enough north to be in the zone of light and erratic rainfall, Australia tends to have poor soils and a resulting low plant productivity. Thus, it cannot support a large population, especially away from the coastal regions. In turn, there are few people, especially trained people, to prospect for vertebrate fossils. In consequence, the Australian record of fossil (terrestrial) vertebrates is tolerably complete from the Pleistocene as far back as the end of the Oligocene (about 25 million years). For times before this, it is incorrigibly spotty: there are a few tetrapods from the Permian and the earliest Triassic; fewer from the early part of the Middle Jurassic; and some dinosaurs, crocodilians, turtles, and mammals from the middle of the Cretaceous (about 95–120 million years ago). If the history of land animals in Australia after the beginning of the Miocene can be thought of as a movie film, these earlier deposits afford nothing but a handful of still frames. The time elapsed since the opening of the Miocene makes up less than ten percent of the history of vertebrates on the land, so there is much that is missing. In quality, this part of the fossil record of terrestrial tetrapods is generally comparable to that known from North America prior to the Civil War, or for Europe at the time of Napoleon (and Cuvier). The Early Tertiary is equally poorly represented, although the turtles are reasonably well known.

It is possible that Australia was as completely overrun by ice sheets during the Permo-Carboniferous glaciation (about 250–350 million years ago) as Antarctica is today. In any case, there are very few tetrapod fossils from pre-Permian Australia. Late Devonian tetrapod trackways are known from Victoria (Fig. 2.2), and a single jaw of a very primitive amphibian (ichthyostegalian) has been found in New South Wales (Fig. 2.3). An Early Carboniferous site from eastern Queensland has yielded remains of other very early amphibians (Fig. 2.4)—colosteids, anthracosaurs, and temnospondyls (Thulborn et al. 1996)—all still under study. This is supplemented by a few temnospondyl specimens from the later Permian of the Sydney Basin in New South Wales (Fig. 2.5). This Paleozoic material is too fragmentary for any enlightening comparison with specimens from the rest of the world. It shows some of what lived in this part of the world, but little else.

Rocks of the Early Triassic are more forthcoming. Deposits in southeastern Queensland, eastern New South Wales (Fig. 2.6), Tasmania, and along the coast of Western Australia have produced a fauna that includes more temnospondyl amphibians (Fig. 2.7) as well as procolophonians. These small reptiles, which would have looked much like stout, stumpy-legged lizards, are now thought to be closely related to the ancestors of turtles (Fig. 2.8). There are also rare fossils of early relatives of the dinosaurs (thecodonts) and mammals (therapsids). The fauna of temnospondyls is diverse (Fig. 2.9) and the procolophonians were apparently common in southeastern Queensland, although absent elsewhere.

There are several endemic genera and species among these animals, but all of the forms seem to be unexceptional Early Triassic examples of their respective groups. They are in no way as unusual—compared to animals from other lands—as are the living Australian mammals. Tony Thulborn of the University of Queensland has argued (1986) that although the individual animals (i.e., taxa) found are unremarkable, the fauna itself is unusual. Temnospondyls were rather more common

Fig. 2.2. A Late Devonian tetrapod trackway from Genoa River, Victoria. (Courtesy of Prof. J. Warren, Monash University.)

Fig. 2.3. (below) The right lower jaw of Metaxygnathus denticulus, a Late Devonian ichthyostegalian amphibian from central-east New South Wales, one of the oldest tetrapod fossils. The jaw is approximately 12 cm long, and so may represent a creature a meter to a meter and a half in length. (From Campbell and Bell 1977.)

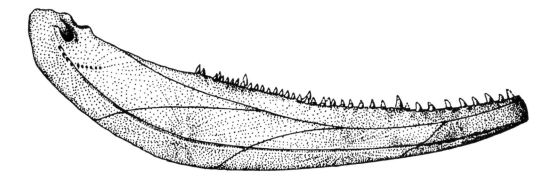

Fig. 2.4. (above) Restoration of a
Carboniferous temnospondyl
amphibian, life size, whose fossils
were found in east-central
Queensland. The photo is of a
temporary display in the
Queensland Museum in 1996.

Fig. 2.5. The fossil of a small
temnospondyl (Blinasaurus
wilkinsoni) from the later
Permian of the Sydney Basin in
New South Wales. This specimen
and exhibit was in the now-
defunct Geological and Mining
Museum in Sydney
(photographed in 1981).

Fig. 2.6. Exposures of the Triassic Bulgo Sandstone along the coast of New South Wales in Boudi State Park north of Sydney. Perhaps for obvious reasons, fossils have been found only in the platforms at sea level.

here than in the faunae of other (contemporaneous) continents (Fig. 2.10). Indeed, some groups of temnospondyls are thought to have originated in what is now Australia (e.g., Warren and Black 1985). Thus the East Gondwanan (Australo-Antarctic) land mass may have been a distinct zoogeographic region at this time. Unfortunately, the record is disquietingly incomplete, and the abundance of temnospondyl fossils may simply reflect a bias toward animals from the habitats where temnospondyls happened to be prevalent, presumably swamps, marshes, and the like. Still, in view of what we see in the later Mesozoic and after, this is a suggestive first glimpse of the evolving Australian fauna.

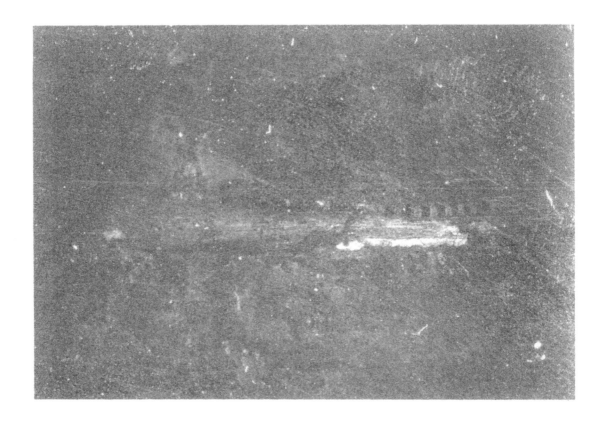

Fig. 2.7. (above) The mandible of
the large temnospondyl amphibian Bulgosuchus gargantua,
photographed before collection.
The anterior part of the jaw is to
the right, and some teeth can be
seen: the posterior part, to the
left, is missing. This specimen
was found on the shore platform
in the region shown in Fig. 2.6.

Fig. 2.8. A sculpture of a restored
procolophonian, based on
material found in the Triassic
beds of Queensland (and,
probably, Procolophon
trigoniceps, of South Africa).
This life-size sculpture was
discovered when the Queensland
Museum moved from its old
building to the new one.

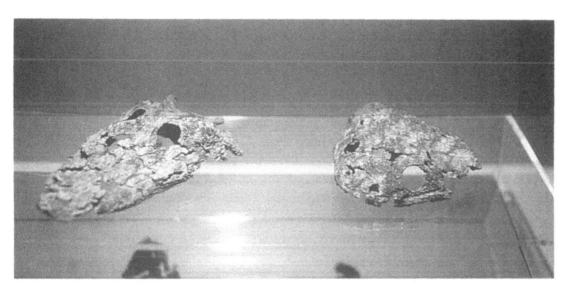

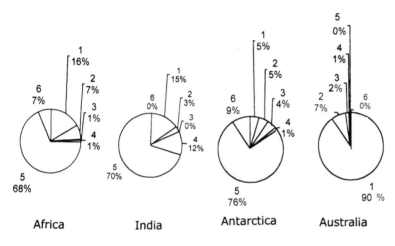

Africa    India    Antarctica    Australia

*Fig. 2.9. (above) The skulls of two temnospondyls from the Triassic of Queensland,* Watsonisuchus aliciae *(left) and* Xenobrachyops allos *(right). These skulls were painstakingly reconstructed from pieces by Dr. Anne Warren of LaTrobe University (Melbourne). The photograph was made during a temporary exhibition at the National Science Museum, Tokyo, in 1998.*

*Fig. 2.10. The proportions of the groups of Early Triassic tetrapods in various parts of Gondwana-land. Key: 1. temnospondyls; 2. procolophonians; 3. small lizard-like reptiles; 4. thecodonts; 5. dicynodonts; and 6. therapsids other than dicynodonts. In Africa, India, and Antarctica, dicynodonts (5) make over 65 percent of the fauna; in Australia they make up less than 0.5 percent (here rounded off to 0%). In Australia temnospondyls (1) make up 90 percent of the fauna, but in the other lands they make up less than 20 percent. (Data from Table 1 of Thulborn 1986.)*

Jurassic tetrapods are represented by a skeleton and a half (and some fragments) from southeastern Queensland and two bones from Western Australia. A sauropod caudal (vertebra from the tail) and the distal part of the tibia of a small theropod named *Ozraptor* show only that dinosaurs inhabited the western part of the continent during the Middle Jurassic (Bajocian). The fact that the dinosaur represented by the tibial fragment was named at all indicates how desperate Aussie paleontologists are for Mesozoic tetrapod specimens. The Queensland material includes a primitive sauropod, *Rhoetosaurus*, represented by the posterior half of a skeleton and the large temnospondyl *Siderops*.

A jaw fragment from one of these amphibians had been found in an isolated block of rock in the bed of the Brisbane River early in the 1940s. At that time it was widely believed, on the basis of evidence as reliable as any in the fossil record, that temnospondyls became extinct with the close of the Triassic. Thus, this fossil was thought to have been

Fig. 2.11. The track of large theropod dinosaur, from Middle Jurassic beds exposed in the Caledonian Colliery (now shut down) outside Brisbane, southeastern Queensland. When photographed (1993), this specimen was in a private collection.

reworked from local Triassic deposits—in short, that a Triassic fossil had been eroded free of the rock, washed into the bed of a watercourse, and reburied in somewhat younger rocks, all early during the Jurassic. Such events are rare, but have happened in some parts of the world.

The discovery of the skeleton of a large temnospondyl in the indisputably Jurassic (Liassic) Evergreen Formation showed that in this part of the world temnospondyls did survive the end of the Triassic (Warren and Hutchinson 1983). Since then, Jurassic temnospondyl fossils have been found at several localities in central, eastern, and southeastern Asia, clearly showing that the fossil record cannot always be taken at face value.

The footprint record for the Jurassic of Australia shows a little more of what kinds of animals lived there whose bones were not preserved. Large and small ornithopods and theropods (Fig. 2.11), or at least the tracks of large and small individuals, and a possible thyreophoran (armored dinosaur) track indicate that these groups were also present.

Cretaceous tetrapods—although effectively known only from Aptian-Cenomanian times (ca. 95–120 million years ago)—are the best known for any time before the Miocene. Moderate to large dinosaurs lived in central Queensland, and isolated bones of smaller dinosaurs have been found in northern New South Wales (Figs. 2.12, 2.13) and southern Victoria. Extensive trackways have been found in central Queensland (Fig. 2.14) and the northwestern coast of Western Austra-

lia. The latter are associated with what seems to be the preserved floor of a forest. Two- or three-meter-long ornithopods are prominent among fossils from Victoria and New South Wales and are also found in southern Queensland. Central Queensland supported a small ankylosaur (*Minmi*), large ornithopods (*Muttaburrasaurus*), and at least one kind of sauropod (*Austrosaurus*). Body fossils from the Winton Formation represent only sauropods among the dinosaurs, although the Winton trackways represent only theropods and ornithopods.

It is possible, although far from certain, that central Queensland had a distinct dinosaurian fauna, different from that of Victoria and New South Wales (and maybe from southern Queensland). Two of the central Queensland dinosaurs, *Muttaburrasaurus* (Fig. 2.15) and *Minmi* (Fig. 2.16), are unusual animals that seem to represent lineages that evolved in isolation in Australia. But *Austrosaurus* has recently been recognized as a titanosaur, a group of sauropods prominent in the Gondwanan continents during the Cretaceous—and by the end of that period having spread into Europe and North America. Four kinds of dinosaurs are known from footprints, one of which might be *Muttaburrasaurus*. The others include a large and a small theropod and a (relatively) small ornithopod that might be one of those found in southeastern Australia.

Other Cretaceous fossils represent turtles, unusual and uniformly rather small crocodilians, and mammals. These latter have received disproportionate attention because of the interest inherent in the ori-

Fig. 2.13. (above) The mail drop at Grawin, another fossil-bearing opal field in New South Wales not far from Glen Garry and Lightning Ridge, in 1982. In addition to the colorful structure, this gives a good impression of the nature and flatness of the country in which the opalized fossils are found in northern New South Wales.

Fig. 2.14. Cleaning the mid-Cretaceous rock surface with dinosaurian tracks near Winton, central Queensland.

gins of the unique modern Australian mammalian fauna. These include the possibly platypus-like monotreme *Steropodon* (Fig. 2.17). Another, *Ausktribosphenos,* has been claimed to be an early placental mammal, a claim so far not generally accepted by the paleontological community. Some paleomammalogists now contend that both forms are monotremes (or closely related) and have evolved the characteristic placental type of dentition (tribosphenic dentition) independently of placental mammals (Stokstad 2001). A single lepidosaur bone—a worn humerus—has been discovered in the Early Cretaceous beds of Victoria (Fig. 3.8). But most intriguing was the discovery of temnospondyls in Early Cretaceous beds in Victoria (Warren, Rich, and Vickers-Rich 1997). This material represents a substantial increase, by about 25 percent, in the known period of time that these animals lived and

*Fig. 2.15. The mounted, reconstructed skeleton of the Early Cretaceous ornithopod* Muttaburrasaurus langdoni, *in the Queensland Museum. This reconstruction is based on the type specimen found in central Queensland. In fact, it is not clear that the animal could have adopted such an upright pose. In the top background, replicas of Cretaceous dinosaurian trackways from Winton (cf. Fig. 2.14), central Queensland, may be seen. (Photograph courtesy of Bruce Cowell and the Queensland Museum.)*

almost certainly represents a relict population surviving about 50 million years after their extinction elsewhere.

For the remainder of the Cretaceous, tetrapod remains are almost nonexistent in Australia, as are exposures of rocks of that age. Two isolated bones alleged to be from theropod dinosaurs—one of them quite possibly deriving from a marine saurian—and a pterosaurian element have been found in Western Australia.

A small fauna of very Late Cretaceous (Campanian-Maastrichtian) dinosaurs is known from North Island, New Zealand, due to the indefatigable efforts of Joan Wiffen. The story of these efforts deserves a book of its own, and fortunately it has one, written by Joan herself

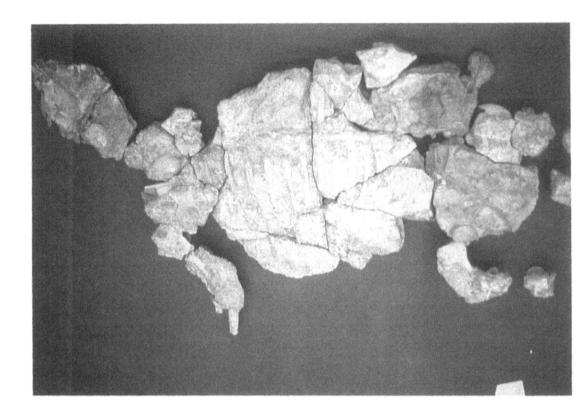

Fig. 2.16. (above) The fossil
skeleton of the Early Cretaceous
ankylosaur Minmi, laid out in the
position in which it was found
but inverted (it was discovered
upside down). The specimen is
seen from directly above and has
been somewhat flattened during
preservation. It was found near
Richmond, north-central
Queensland in 1990.

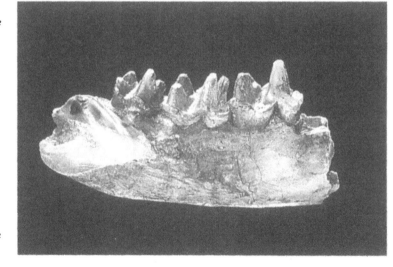

Fig. 2.17. The jaw of the Early
Cretaceous monotreme
Steropodon galmani preserved as
transparent opal, found near
Lightning Ridge, New South
Wales. The specimen is but a few
centimeters long. After more than
a century of speculation, this
specimen finally demonstrated
that monotremes were, indeed,
ancient mammals. (Photograph
courtesy of Alec Ritchie and the
Australian Museum.)

*Fig. 2.18. A probably theropod caudal vertebra from the Late Cretaceous of North Island, New Zealand. This specimen, discovered by Joan Wiffen, was the first clear evidence that dinosaurs had inhabited the land that is now New Zealand.*

(Wiffen 1991). She took up paleontology as a hobby after having retired from farming. Almost single-handedly, but with the help of her late husband and some friends, she discovered, excavated, and prepared a fauna of Cretaceous reptiles. Having taught herself how to remove the fossils from the rock and clean them, she then went on to teach herself how to write and publish the scientific papers describing them. It is often said that science is "democratic" in that anyone with the interest (and the time) can learn to be a scientist, but rarely is this demonstrated more clearly than in Joan Wiffen's case. Her discoveries include one or more theropods (Fig. 2.18), a small ornithopod, an ankylosaur (also small), a large titanosaurian sauropod, and a pterosaur—all, unfortunately, known from very incomplete specimens. Contemporaneous bird fossils from South Island remain unstudied. These animals seem to have lived under relatively low temperatures with a mean annual temperature near 15°C.

Similar, but more extreme claims—a mean annual temperature around or just below 0°C—have been made for the Early Cretaceous climate of southern Victoria. The evidence cited is consistent with this interpretation, but each item of evidence is also consistent with other interpretations. For example, the intermixing of masses of sediments of different types has been interpreted as evidence of the overturning of unconsolidated sediments during the summer melting of permafrost. However, such intermixing can also occur during earthquakes, which can mobilize unconsolidated sediments, producing sand boils and sand volcanoes. These structures also can mix together masses of sediments of different types. Furthermore, some seismic activity must have been associated with the protracted separation of Australia from Antarctica. The Victorian fossil plants suggest temperatures more like those of Late Cretaceous New Zealand, and forests are known to have been growing further south in Antarctica around this time, their preserved roots providing clear evidence that permafrost was absent. Both Victoria and North Island were probably cool during the Cretaceous, since both were near the South Polar Circle, but more convincing evidence is needed before accepting periglacial dinosaurs in Victoria. Recent geochemical work (Kerr 2000b) suggests that this part of the Early Cretaceous was the warmest time during the past 150 million years, which is difficult to reconcile with the claim of seasonal freezing temperatures for southern Australia.

Excepting Antarctica, no other continent has a Mesozoic terrestrial fauna that can be adequately mentioned in so few pages: for the Early Cenozoic, the situation is no better. Although the Tertiary is usually divided into five epochs, a division into two periods is useful. These are the Paleogene, from the beginning (65 million years ago) to about 24 million years ago, and the Neogene, which lasted until the beginning of the Pleistocene. Two Australian Paleogene sites are worth noting, both in southeastern Queensland. Both are probably Eocene in age, although Murgon could be as late as Oligocene, and both could perhaps be as old as Paleocene.

Redbank Plains, a developing suburb of Brisbane, is built over the deposits of a Paleogene lake. The vertebrate fossils are hollow natural molds in ironstone concretions, most or all of the bone having been destroyed by weathering. The fauna has not been systematically studied, but it includes insects, bony fish, and chelid turtles. The foot of a large, probably ground-dwelling bird was recovered—ironically not by a vertebrate paleontologist but by New Zealand paleobotanist Mike Pole. There is no trace of mammalian or crocodilian fossils. Not finding mammals in the middle of a large pond or lake is not especially surprising, but crocodilians were expected since they occur elsewhere in Queensland at (more or less) this time.

The other site, near the small town of Murgon, does produce both mammals and crocodilians, along with birds, turtles, and lizards. The material is noteworthy in two respects. First, the tooth of a condylarth (*Tingamarra*) has been reported, a possible placental (or eutherian) mammal, by the far oldest known from Australia (unless *Ausktribosphenos* is one). This tooth may indicate that eutherian mammals did live in this part of the world and that they may have become extinct in competition with marsupials. This is particularly interesting in view of the flexibility of the marsupial reproductive strategy in the face of unpredictable climate and rainfall and the increasing aridity. As will be discussed later (chapter 6), marsupial reproduction permits the female to control how much energy and effort she puts into raising a young (Parker 1977; Dawson 1983). She may terminate this (thus killing the offspring) during harsh times such as droughts. In contrast, a female placental mammal lacks this ability, and in difficult times she runs the risk that both she and her young may die. With the increasing harshness of its climate as Australia moved north, the marsupial strategy would seem to have been the more effective. However, this harsher climate appeared well after the fossils of Murgon were living animals, so one wonders why eutherian fossils were not more common here (and elsewhere) if they were in fact present. We should also note that teeth alone are not entirely convincing as evidence of a eutherian mammal, especially from an ill-defined group such as condylarths; thus, *Tingamarra* may not belong to this group at all. Many in the vertebrate paleontological community prefer to reserve judgment on this animal until more convincing evidence is found.

A less contentious and perhaps under-appreciated find is the discovery of marsupial teeth resembling those of the South American microbiotheres. Judging from the single surviving species, *Dromiciops australis* of southern Chile, microbiotheres were small, arboreal, opossum-like marsupials that have been proposed as close relatives of the Australian marsupials. Their occurrence at Murgon lends credence to this idea.

A Late Oligocene site (about 25 million years old) at Geilston Bay in Tasmania has produced a number of marsupials. With this site the quality of the vertebrate fossil record dramatically improves, both in the number of localities and in the quality and completeness of the fossils recovered. In Australia, the transition from the Paleogene to the Neogene marks the beginning of a serious fossil record for land-dwelling vertebrates. Sites at Riversleigh in northwestern Queensland and in the central Australian desert have yielded a wide variety of Oligocene and Miocene tetrapods, those from Riversleigh being exceptionally well preserved. Monotremes (re)appear in the record at Riversleigh with the platypus *Obdurodon*. As far back as can be discerned, monotremes seem to have included only platypuses and echidnas as they do today (although the giant echidnas have become extinct). By the time that the oldest deposits at Riversleigh were laid down, the modern groups of marsupials were already established, implying that those groups probably arose earlier in the Oligocene. Families of marsupials now represented by one or a few genera, such as koalas and wombats, were notably more diverse; and there were other groups, now extinct, such as the possum-like ektopodonts and the still-mysterious yalkapariodonts. These groups, unlike many others, seem to have become extinct well before the end of the Pleistocene. At any one time, fossil marsupials show a substantially greater diversity at the generic level than they do today, but less so at higher levels.

The records for other groups—birds, lepidosaurs, crocodilians, and frogs—also drastically improve (although much of the bird material has yet to be studied) near the beginning of the Miocene. The main difference between the bird faunae then and those of today was the presence of the large flightless dromornithids. Chelid turtles, whose fossil record effectively begins at Redbank Plains, seem to have already evolved modern genera by that time, and (on the basis of studies so far) show no indication of a greater diversity in the past than now. Chelids are the most common turtles of the Southern Hemisphere, and are found also in Africa and South America, but there were also other turtles in Australia at this time. Trionychids (soft-shelled turtles) and the horned meiolaniids were present but became casualties of the Pleistocene extinctions. The lepidosaur record is not the best, but as with chelids, there is no sign of substantially greater diversity in the past. Australia is host to quite a diversity of squamates now, and the major casualty was *Megalania*.

Crocodilians are a different story. Prior to the arrival of the Indo-pacific crocodile, *Crocodylus porosus,* Australia was inhabited by a diversity of crocodilians that completely disappeared by the end of the Pleistocene. Some were probably replaced by *Crocodylus porosus* and the Freshwater croc (*Crocodylus johnstoni*), but others seem not to have been replaced at all. The Pleistocene saw the extinction not only of *Megalania* but also of these endemic southwestern Pacific crocodilians, the mekosuchines. These animals seem to represent a Gondwanan lineage that radiated in Australasia. First recognized by Christoph Balouet and Eric Buffetaut in Paris, from New Caledonian fossils (1987), they were shown by Paul M. A. Willis, then a student at the University of New South Wales, to have been prominent in the Australian Cenozoic. They have also been found in Fiji, Vanuatu, and New Zealand, and perhaps in New Guinea as well. But the material so far obtained from the latter island is too fragmentary to be sure. In Australia they adopted a variety of forms, some of them thought to have been terrestrial, as opposed to amphibious like modern crocs. One of these probably terrestrial animals, *Quinkana,* is mainly represented by isolated teeth from northeastern Australia (Queensland and northern New South Wales).

## Ice Ages Without Ice: Pleistocene Australia

During the Cenozoic, Australia drifted slowly northward from a position that conferred on it a cool, seasonal climate into a warmer but drier climatic region. During much of this time the global climate itself cooled. This cooling was to some extent offset in Australia by the northward drift, but with the cooling we might expect an increase in aridity (due to a reduction in evaporation following the cooling of the oceanic surface waters), which would have added to that caused by Australia's movement north. Two major influences in Australian Cenozoic history were related to this cooling: the establishment of the circum-Antarctic oceanic currents (Fig. 2.19), and the glaciation of Antarctica that followed. After the separation of both Australia and southern South America from Antarctica, cold currents in the southern ocean could circulate entirely around that continent without being deflected northward (and hence warmed). This probably occurred during the Oligocene, and the accumulation of cold water in the southern ocean both cooled the Australian climate and contributed to its aridity.

Central Australia was, at least in part, both forested and well watered during the Early Miocene, about 20 million years ago. Temperate rain forests were dominant in the southeast, and rain forests were found in the north. Pollen assemblages suggest that grasslands spread in central Australia during the Middle Miocene, probably through the

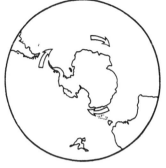

Late Eocene                    Early Oligocene

*Fig. 2.19. The establishment of the circum-Antarctic current. During the Late Eocene, although currents could freely pass between South America (SA) and Antarctica (An), they were blocked (black arrow) from passing south of Australia (Au) by the South Tasman Rise (ST, dotted). During the Oligocene, the Rise drifted north and opened a passage for cold and deep currents to circulate around Antarctica (white arrows), thus permitting chilling of the continent. NZ, New Zealand. (Modified after Lawver et al. 1992.)*

plains between the riverine gallery forests. Sedimentary rocks from this period suggest that the river channels were not strongly developed and thus that the countryside was flat, with at most only low relief—as much of Australia still is today.

Australia experienced some physical changes similar to those that occurred elsewhere in the Pleistocene. There were at least three successive ice sheets in Tasmania, as well as at least one in the Snowy Mountains of New South Wales and Victoria (in both this region and Tasmania evidence of earlier glaciations may yet be found). But these were basically enlarged mountain glaciers, not the kilometers-thick sheets that occurred in the Northern Hemisphere. As with Torres Strait (separating Cape York, Australia, from the southern coastal lowlands of New Guinea) in the north, drops in the sea level repeatedly joined Tasmania to the mainland. Bass Strait (separating Australia from Tasmania) is deeper than Torres Strait, and so would have remained in existence for a longer period of time; thus, Tasmania was isolated from Australia longer than Australia was from New Guinea. Dunes and grasslands formed on the emergent floor of Bass Strait, and this region may have formed a barrier to migration to Tasmania even when the sea was absent.

In some regions the slope of the continental margins is so slight that the estimated rate of sea level rise of 25 mm/year after the Weichselian glaciation could have caused some shorelines to retreat inland at 10–30 meters/year (Galloway and Kemp 1984). There were periods of greater and lesser rainfall. At the last glacial maximum, Lake George in the mountains of New South Wales received approximately half as much precipitation as it does now. During these arid times, dunes formed in central Australia. Lakes and wetlands expanded or dried with the changing climates. Falls in sea level probably contributed to the aridity, since they would increase the surface area of the continent, thus reducing the oceanic area available for evaporation in this region, and alter the flow of oceanic currents, particularly in the tropical north.

Australians are pleased to think of their homeland as the "lucky country," and compared to events in other parts of the world at this time, this is an apt phrase for Pleistocene Australia. Australia was not subject to the extent of glaciation that occurred in North America and

Europe. Neither was (nor is) there extensive volcanism as there is in Indonesia, for example. Desirable as this may have made Australia as a Pleistocene home, the absence of these influences had less felicitous results in the long run. Events that left easily datable traces in Pleistocene strata elsewhere did nothing of the kind in Australia. (Another of the adverse consequences of Australia's "quiet" tectonics is discussed in chapter 6.) In consequence, Pleistocene chronology and correlation is almost nonexistent: most Australian Pleistocene sites can only be recognized as Pleistocene, with no clue as to how recent or ancient within that epoch they might be.

When this chapter was originally composed, I wrote: "So although we know that large tetrapods died out sometime during the Pleistocene, and roughly when, we don't know if this occurred abruptly, in a stepwise fashion (some first and others later) or gradually; we don't know if it occurred first in the arid 'Centre' or in the more humid coastal regions; we don't know if it occurred later in the tropical north or in the temperate south or at the same time in both." But this has now changed. Richard Roberts at the University of Melbourne, together with Tim Flannery (about whom more later) and other colleagues, have measured dates for beds laid down at about the time of the megafaunal extinction (Roberts et al. 2001). The extinction occurred abruptly and at much the same time in the arid Centre and the coastal regions. Sites from the tropical north (and northern and western interior desert) were not included; and from my experience with some of these sites, I suspect the conditions were not appropriate for obtaining a reliable date.

During the Pleistocene—as before and as now—Australia was an isolated island continent. It was, presumably, repeatedly in contact with New Guinea through a vast but low and probably well-watered (at least in parts) plain, now submerged. The joint land mass is termed Meganesia (Fig. 2.20). It had been possible for some time, probably since at least the Early Pliocene, for tetrapods to migrate from Asia via Indonesia into Australia, and some lizards and rodents and the Indopacific croc did so. It was also possible for tetrapods to emigrate from Australia into Asia, but remarkably few (apparently only marsupials) did. They reached only as far as Sulawesi but seemingly failed to become established anywhere further west on the islands that, during the glacials, became part of the Asian mainland (Fig. 2.21).

## Cenozoic History of Land Plants in Australia

A comprehensive history of the flora of Tertiary Australia is yet to be written. Robert Lange, a botanist at Adelaide University in South Australia, characterized what is known of the history as presenting two distinct stories, depending on what kind of evidence was consulted (Lange 1982). Phylogenetic and chromosomal studies, largely of modern plants, indicate that plants characteristic of heath and eucalypt forest were widespread but became separated into eastern and western regions by some ecological barrier (the nature of which remains unknown) during the Miocene. On the other hand, studies based on pollen assemblages indicate that widespread rain forest, including

Fig. 2.20. Meganesia, the continent that incorporated Australia, Tasmania, and New Guinea during low stands of the sea in the Pleistocene. Considerable areas of what is now continental shelf were then dry land. (From maps in Allen and O'Connell 1995.)

Fig. 2.21. (below) Although some animals from Asia emigrated to Australia, Australian animals did not invade much into Asia. The heavy dashed line indicates how far diprotodont marsupials (mostly possums) got into Indonesia, and the light dashed line the limit of polyprotodont marsupials (mostly bandicoots). Heavy continuous lines indicate the low sea stands during the Pleistocene, dotted lines modern sea level. No marsupials established themselves on any lands that were continuous with the Asian mainland during low sea stands. (Modified after Raven 1935.)

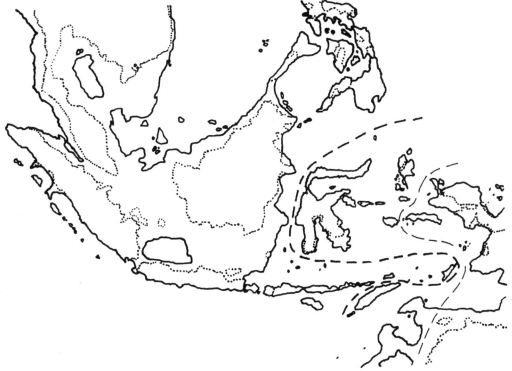

*Fig. 2.22. A eucalypt on the banks of the Mitchell River, northeastern Queensland.*

southern beech (*Notofagus*) and conifers, persisted almost to the Pleistocene. And then the rain forest was replaced by other forests, not more open plant communities. One interpretation of this seemingly inconsistent record is that the primeval Early Tertiary Australian rain forest was replaced for certain periods and in certain regions by other plant communities in response to the increasing aridity (Singh 1982). In other words, the rain forest (when present) contributed the pollen, but the other plant communities contributed the modern plants used in the chromosomal and other studies. Thus, each source of evidence tells the history of only some components of the Australian flora, not of the flora in its entirety. As the climate fluctuated, these different communities formed and vanished, and as the dryness became more prevalent, the rain forest contracted until finally it was almost lost.

Toward the end of the Miocene, the temperature fell slowly and then declined sharply in the Early Pliocene. As the Antarctic ice sheet expanded, the Australian climate probably became drier and rain forests shrank. Sometime during the Pliocene, the rain forests of the interior had completely disintegrated and become replaced by trees suited to drier climates, such as gums (*Eucalyptus*) and she-oaks (*Casuarina*), and by grasses. Also during this period the already-existing Great Dividing Range of eastern Australia was again elevated, as were the mountains of New Zealand and New Guinea. The latter provided a diversity of high-elevation habitats in the warm equatorial north for creatures and vegetation adapted to the cool climate of Early Tertiary

*Fig. 2.23. A fossilized gymno-sperm cone, probably from an araucarian, from the Early Cretaceous of Lightning Ridge, New South Wales. This specimen was in the now-closed Geological and Mining Museum, Sydney.*

Australia. The cool Pliocene climate was followed by a warm and presumably moist period during the Early Pleistocene that did not last long. There is evidence that rain forest returned (temporarily) to western New South Wales at this time. The climate, however, cooled again.

Eucalypts, now so characteristic of Australia (Fig. 2.22), became the dominant woodland trees only during the latter part of the Würm (Singh 1982). The lost forests of Australia were made up of other (angiosperm) trees together with the stately conifer *Araucaria* (Fig. 2.23). During this time, about 38,000 years ago, the pollen records show that the complexion of the plant communities changed, with a decrease in the rain forest vegetation and an increase in the eucalypt-acacia (sclerophyl) woodland. This is attributed to the effect of fires set by people. It is generally accepted that the open, park-like character of the Australian landscape that the English found so inviting as well as the dominance of eucalypts were largely due to the influence of the human inhabitants and their fires.

## Pleistocene Vertebrates in Australia

During the Cenozoic, Australia crept northward from a position near the Antarctic Circle in the Cretaceous (approximately straddling 50° S) to its present position straddling the Tropic of Capricorn (approximately 24° S). As the global climate cooled during the Cenozoic, Australia continued its equatorward slide, thus to some extent mitigating the effects of the cooling. But some are won and some are lost: as Australia moved north, it left the region that meteorologists call the subpolar low (in reference to the general air pressure), in which the air tends to rise, to enter the subtropical convergence, a region where it tends to fall. As air rises, it cools, its moisture condenses, and rain (and sometimes snow) falls; as air falls, it warms, and its moisture evaporates. Since most of the moisture of the sinking air would have precipi-

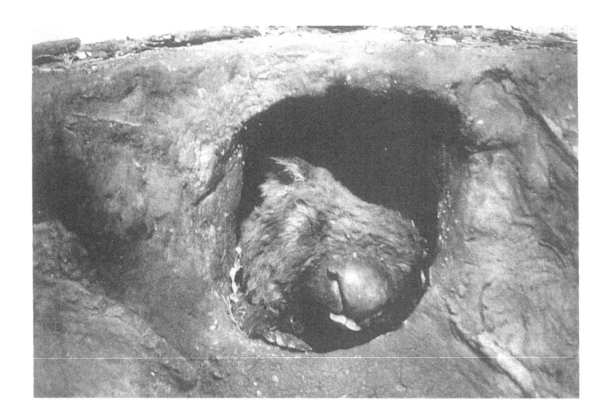

Fig. 2.24. The exhibit of the giant Pleistocene wombat, Phascolonus gigas, *asleep (a typical "activity" of wombats) at the mouth of its burrow in the Queensland Museum, 1990.*

tated out when it initially rose, the air becomes dry, and any water on the ground tends to evaporate. Thus, Australia crept from a humid climatic zone into an arid one, and more and more of the continent became increasingly dry as the Tertiary progressed.

At Riversleigh, in the north, rain forests seem to have persisted at least through the Miocene (Archer, Hand, and Godthelp 1991), so the weather was humid enough to support them until perhaps five million years ago. But the glacial cycles that commenced in the Pliocene, which began at that time, also brought dry interpluvials, and aridity increased. This turned the remaining moist regions near the coasts into refuges for plants and animals driven out of the drying central regions, which now occupied about 70–80 percent of the continental area. These refuges were regions in which the weather conditions of earlier times still persisted, where the plants and creatures that required such conditions could still survive.

Compared to older faunae from Australia, that of the Pleistocene is well known, but compared to those from other lands it is not. In addition to modern forms—all of which presumably date back at least into the Late Pleistocene—there was a variety of creatures that are now extinct. Then as now, there were kangaroos, koalas, wombats, possums (both phalangerids and pseudocheirids), bandicoots, dasyures, and other marsupials, as well as platypuses, echidnas, bats, and rodents.

In addition there were the diprotodontids, *Diprotodon* and its kin, *Zygomaturus,* and *Euryzygoma,* large, superficially wombat-like ani-

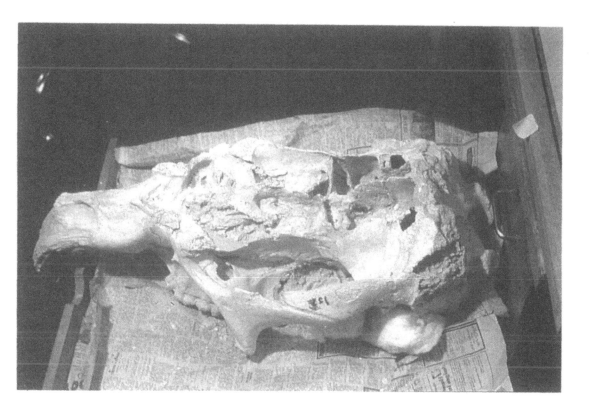

Fig. 2.25. A skull of the *Pleistocene marsupial* Diprotodon, *found at Bacchus Marsh, Victoria, now in the Museum of Victoria. The skull is seen from the left, looking down obliquely from above. The roof of the skull is missing, so that the extensive sinus chambers filling much of the skull and surrounding the braincase can be seen.*

mals, often called "giant wombats" in Australia. Because of this name, they are confused with the true giant wombats (*Phascolonus*) of the Pleistocene, but wombats are more closely related to koalas than to diprotodontids. Giant wombats were about twice as large as living ones—large enough, if they excavated burrows at all, to have dug out passages almost the size of small subway tunnels (Fig. 2.24).

Little is known of how diprotodontids lived. Their skull is large, but much of what appears to be braincase actually houses extensive nasal sinuses (Fig. 2.25), and the brain itself was substantially smaller than would be supposed from looking at the outside. The nasal bones of the snout are elevated and placed somewhat back from the tip, unlike those of any other known beasts. Perhaps they supported an extensible, mobile snout, a kind of short trunk like those of tapirs. Or perhaps it was something else altogether that no one has imagined. Both wrists and ankles have ball-and-socket joints, reminiscent of those of a lunar lander, on which large, flat (plantigrade) fore and hind feet were accommodated. The ankles of ground sloths have a similar structure. Unfortunately, ground sloths are also extinct, so we can't observe how they move about. But however diprotodontids lived, it seems to have been successful, because judging from the amount of wear on their teeth, most Pleistocene diprotodontids died at an advanced, sometimes very advanced, age (Fig. 2.26). In many specimens the teeth are so worn down that, in the living animal, they can hardly have projected above the gums. This contrasts sharply with Pleistocene kangaroo jaws, which

*Fig. 2.26. A jaw of* Diprotodon *from Late Pleistocene deposits of Darling Downs, southeastern Queensland. The teeth, particularly toward the front (which is to the left) are heavily worn. This indicates that the animal lived to be quite old (at least for a diprotodont). The more posterior teeth erupt later in life than the more anterior, so the teeth become less worn toward the back of the jaws (to the right).*

often have unworn teeth, suggesting that relatively few of these marsupials lived to old age.

The macropod lineage today is made up of the larger kangaroos and wallabies and the smaller rat-kangaroos, potoroos and bettongs. All of these forms have been found as fossils, but a major group of macropods did not survive the Pleistocene. These are the sthenurine kangaroos, large forms with relatively short, deep skulls and a hind foot "reduced" to a single, large toe. They are thought to have browsed on trees, seizing the branches with their long, mobile forelimbs. Although this lineage is generally considered to be extinct, Tim Flannery, now director of the South Australian Museum in Adelaide, believes that one wallaby, *Lagostrophus,* is actually a surviving sthenurine.

There was also a meat-eating lineage, the propleopine kangaroos, that will be discussed in chapter 6. Thylacolions (*Thylacoleo*) were another group of (leopard-size) carnivorous marsupials that will also be discussed there.

Finally, we must mention the palorchestids. Originally taken to be giant kangaroos (hence their name, indicating "ancient leaper"), they are now regarded as having been more like marsupial tapirs. In the snout, the nasals are even more strongly retracted than those of the diprotodontids, and this is thought to indicate that they had a prehensile snout or even a trunk. Both fore and hind feet held large, deep claws, presumably for digging—although we don't know what they were digging for.

Among the other animals of this time were giant echidnas, giant tortoises, and uncomfortably large crocodilians. *Ninjemys* and *Meiolania* were large tortoises with horned skulls and spiked rings of bone ensheathing their tails to form clubs (Fig. 2.27). With horns and shells and tail clubs, one could reasonably expect these to be abundant fossils, but they are not. *Ninjemys* is represented by a single skull and a tail club, and *Meiolania* by isolated horn cores and a few vertebrae (al-

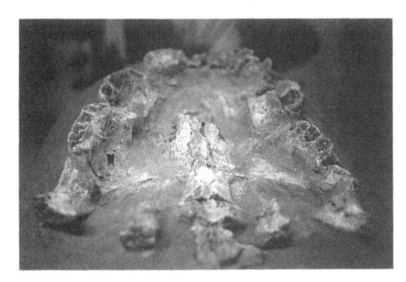

Fig. 2.27. (above) A life-size restoration of the giant Pleistocene horned tortoise Ninjemys oweni, *created by Paul Stumkat (also shown). Fossils of these large, well-armored beasts are unaccountably rare. This formed part of the vertebrate paleontology display of the Queensland Museum.*

Fig. 2.28. *Skull of* Pallimnarchus gracilis, *a large Plio-Pleistocene crocodilian. This specimen, seen from below (i.e., palatal view) and behind, was found in the Dividing Range in east-central Queensland. Originally in a private collection, the skull is now in the Mirani Museum. The posterior part of the skull was not recovered.*

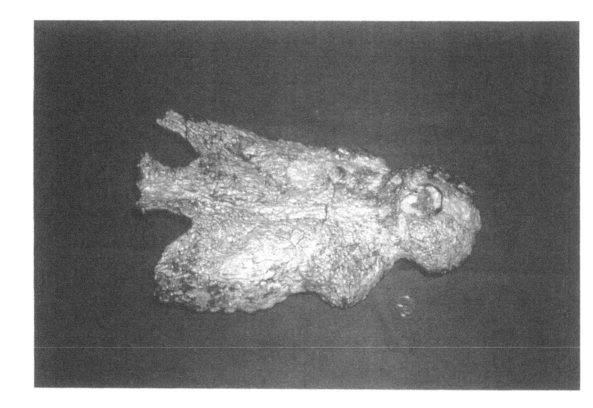

Fig. 2.29. *The snout of* Pallimnarchus, *seen from above. This snout, also of* P. gracilis, *was found in Pleistocene deposits in east-central Queensland. Like that in Fig. 2.28, the back part of the skull was not preserved. (Photograph courtesy of J. Hope.)*

though more complete fossils have been found on islands off the eastern coast). There are enough fossils to show that they really did exist, but little else.

As previously mentioned, the Indopacific crocodile was already in Australia by the Pleistocene, but it shared the continent with *Pallimnarchus,* an equally large mekosuchian form with a broad, very flat skull (Figs. 2.28 and 2.29). This animal was probably an aquatic ambush predator, although one species (*P. gracilis*) had flattened, projecting teeth at the front of the lower jaw that probably indicate some interesting and unusual behavior—if only we knew what. There were other crocs, but we will defer discussion of them until chapter 6.

Throughout the known Cenozoic history of Australia—probably a little less than half of the total—there were no major disruptions to the evolution of animals, despite the changing climate. Of course, the terminal Eocene extinctions seen in the Northern Hemisphere occurred before the well-known part of the Australian fossil record. And there seem to have been some extinctions during the Pliocene (Tedford 1994), but these were minor affairs. Instead, the major extinction came relatively recently, about 46,000 years ago (Miller et al. 1999; Roberts et al. 2001). Until the work of Roberts and his colleagues, there had been no systematic study of the Pleistocene extinctions in Meganesia. One of the reasons for this was the poor understanding of Australian Pleistocene chronology. It is still poorly understood, since the aim of the work cited was to date the megafaunal extinction, not to provide a timescale for

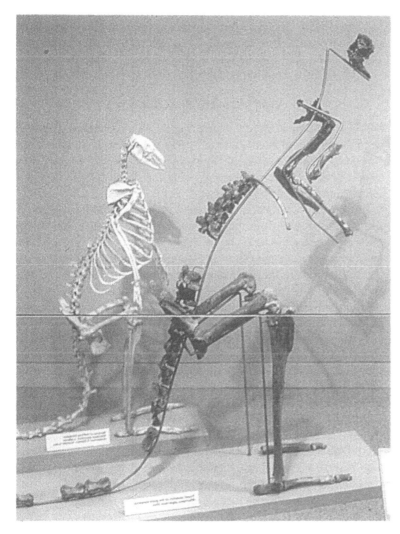

Fig. 2.30. The skeleton of a modern kangaroo (probably Macropus giganteus), compared with that of a close Pleistocene relative, Macropus titan. The latter likely stood almost two meters tall, and so is one example of the greater size of Pleistocene marsupials than their living close relatives. This display was in the Queensland Museum in 1990.

the Pleistocene in Australia, but this work does provide a sorely needed starting point for such a chronology.

The Australian megafauna included large marsupials but not animals as large as those found on other continents. For example, there were no Australian marsupials as large as modern elephants or mammoths. The largest, *Diprotodon,* was about as big as a large horse, although probably heavier. Few, if any, of these animals would have been included in the megafauna had they lived elsewhere in the world, yet for Australia it was the "mega fauna." Among the mammals, as among the lepidosaurs and turtles (as we shall see in the next chapter), the largest members of each group disappeared (Fig. 2.30). This includes the largest kangaroos, as well as large possums (from the Early Pleistocene, perhaps one million years old), echidnas, Tasmanian "devils" (*Sarcophilus*), and wombats. All of the large carnivorous kinds (e.g., thylacoleonids and propleopines) became extinct. Some lineages (giant echidnas, giant wombats, and sthenurine kangaroos) clearly died out, but others ("devils" and some large macropodine kangaroos) underwent a reduction in body size (Main 1978). Although many smaller species died out as well, for marsupials, as for monotremes, birds, turtles, and lepidosaurs, all of the largest animals disappeared.

## Australia and New Guinea

With the falls and rises in sea level, Torres Strait disappeared and re-formed an unknown number of times. There are some obvious and substantial differences between the florae and faunae of Australia and New Guinea. If these lands were recently and repeatedly joined, shouldn't there be more similarity?

Although in science one hesitates to ascribe such differences to mere coincidence, that seems to be the case here. What is now Torres Strait lies near the boundary between the subtropical convergence zone and the equatorial zone, where a humid climate again prevails. Thus, New Guinea retains those forms adapted to humidity, to a mesic (moist) environment; whereas Australia—by and large, and excepting the small rain forests of the eastern coast—is a dry land and so retains animals and plants adapted to a xeric (arid) environment. When the Strait was gone, the emergent lands would have constituted a transitional region between these two different environments. (Taking into account that the global climate would, in general, have been more arid during the glacial episodes, the boundary region likely would have been shifted further north to New Guinea, and the emergent land is thought to have been arid.) The coincidence is that the low-lying lands that flooded to form the Strait lay at this ecological boundary. Thus, both the flooding and the difference in habitats permitted those organisms that ranged from New Guinea through Australia to diverge, adapting to moister conditions in the north and to drier in the south.

In New Guinea itself, during glacial times the region of rain forest was reduced. The high-altitude alpine grasslands had expanded, as had the savanna woodland near (modern) sea level (Flannery 1994). Although the fossil faunae are not well known, they do not seem to represent simply a northern outpost of the Australian faunae. Pliocene remains from lake deposits dating from about 3 million years ago include small diprotodontans, kangaroos, a thylacine, a small dasyure, a rat or mouse, a small cassowary, crocodiles, a snake (constrictor), and a turtle. The Pleistocene specimens seem to date from about 50,000 to 25,000 years ago from the Würm (Flannery 1994). These also include small diprotodontans, kangaroos (including kin of the tree-kangaroos), thylacines, cassowaries, and crocodiles. Missing are the sthenurine kangaroos, large diprotodontans, thylacolions, wombats, and monitors. The diprotodontans all seem to be related, forming a single lineage; and the kangaroos—except for the relatively recent tree-kangaroos—seem to belong to only two lineages.

Of the Pleistocene forms, the diprotodontid *Hulitherium* and two species of the kangaroo *Protemnodon* lived in the montane forests. Others, including the diprotodontan *Maokopia* and a third species of *Protemnodon,* inhabited the subalpine grasslands. The kangaroos are all browsing, leaf-eating forms, with none of the grazing forms common in Australia. The apparent absence of large predators, both marsupial and varanid, may be due to the small size of the plant-eaters. The largest, *Hulitherium,* was estimated to have weighed less than 100 kg (Flannery 1994). Those in Australia were probably more than ten times this weight.

But there is another possibility. We really have no idea what was living in the lowlands, near sea level. The fauna of Pleistocene (upland) New Guinea appears to be distinct from that of Australia, despite being derived from there. A fauna more like that of Australia (perhaps with *Megalania*) that simply didn't exist in the New Guinean uplands may have lived in the now-submerged lowlands. The single date available for megafauna in New Guinea (Kelangurr Cave in Irian Jaya) indicates the fauna there survived at least to about 16,000 years ago, much later than that of Australia (Roberts et al. 2001).

## Australia and Indonesia

Tim Flannery has recently become famous (or notorious) for his 1994 book *The Future Eaters.* In this, he presents evidence that after approximately the mid-Cenozoic, the evolution of the Australian fauna and flora was *ecologically* determined—by levels of available nutrients—as opposed to being *historically* determined—solely by the kinds of animals and plants present in Australia and their interactions. This book met with mixed response for many reasons. Ecological activists admired it because it provided support for their position; most Australians disregarded it because it conflicted with the national dream of transforming Australia into a giant English countryside in the perpetual summer of the southern subtropics (and, incidentally, into the major world power); and some scientists disagreed with it because of its faulty scholarship. There is some faulty, or at least controversial, scholarship, but this relates to peripheral issues and does not affect the major thesis.

Flannery's argument is that Australia has largely "escaped" the geological traumas of continental ice sheets and major volcanism (and we shall return to this in more detail in chapter 6). It is these events, however—together with mountain building—that provide rich and fertile soils for plants. Because the soils of Australia were subject to intense weathering through much of the Mesozoic and all of the Cenozoic, they have now lost most of their nutrients. New fertile soils did not form, since by and large the processes that form them did not occur in Australia. This, together with the increasing aridity and unpredictability of the climate, has limited the number of animals the continent can support. Flannery suggests that in important ways Australia's ecosystems are different from those of the northern continents where most ecologists live, and he provides a plausible account for why they are different.

Asia, source of the creatures of Indonesia as well as of the Australian monitors, is of course one of these northern lands with their different faunae and different ecosystems. The southeastern extremity of Asia, Indonesia, is one name for, geologically speaking, a multitude of places. In biogeography Indonesia is known for Wallace's line,

Fig. 2.31. Wallace's and other
lines of varying biogeographical
significance drawn in Indonesia.
Key: continuous double line,
Wallace's line; long-dashed line,
Sclater's line (based on mammals); short-dashed line, Weber's
line (based on freshwater fish);
dashed and dotted line,
Lydekker's line (based on the
continental slope). Wallace's line
has two "modifications" shown
here: the continuous line
extending to the north between
Kalimantan and the Philippines,
Huxley's line (based on birds);
and that extending to the
southeast of Wallace's line,
Muller's line (based on climate,
and which predates Wallace's
line)—both lines are otherwise
identical to Wallace's line. The
confusing appearance is inten-
tional. (From van Oosterzee
1997.)

proposed by Alfred Russell Wallace, passing between Borneo and
Sulawesi and continuing south from there between Bali and Lombok.
To the north this line extends northeast to pass just south of Mindanao
(in the Philippines) but north of small Palmas Island. Once drawn,
Wallace's line precipitated a host of similar lines drawn and redrawn by
biogeographers, all delimiting the range of some group of organisms to
someone's satisfaction (Fig. 2.31). Finally, Harvard University's George
Gaylord Simpson (1977) complained that such a suite of lines could
have no real meaning.

However, it is not quite as bad as that. Biogeographically, Wallace's
line separates the ranges of many mammals that derive from continental Asia from those that derive from Australia (or Meganesia). Geologically, it separates the continental islands of Asia (to the west)—that in
times of lower sea stand are joined to the Asian mainland—from those
(to the east) that have never been conjoined to Asia. Some of these latter
islands are fragments of the continental plate of Australia and have
drifted north and west at least since the Jurassic, eventually to almost
contact Asia. Their history presumably accounts for the distribution of
their inhabitants by vicariance, that is, the movement of organisms
resulting from the movement of the land on which they dwell. Certainly
some, such as the rodents, have themselves migrated as well—which
confuses the picture, but not beyond understanding.

As with Meganesia, large parts of southeast Asia are now submerged that during glacial times were emergent. In fact, this part of the
world probably experienced the greatest submergence with the postgla-

cial sea rises, with the possible exception of the Chinese coast south from Korea to Indochina. So the environmental and faunal history—and quite possibly even the human history—were significantly affected by the changes in sea level.

The recent (Plio-Pleistocene) fossil record of Indonesia is more complete than that of New Guinea, due in part to the efforts of the searchers for early human fossils. Pliocene deposits are known from Sulawesi, and extensive Pleistocene faunae from Java. But Pleistocene fossils have also been found on Flores, Timor, Sumba, and Sangihe, as well as on Luzon and Mindanao in the Philippines. The Pliocene of Sulawesi yielded fossils of small stegodont elephants, boars, buffalo, tortoises, and crocodiles, but—so far—no monitors. The Pleistocene Javan faunae are much more extensive, and some of these presumably date from times when Java was part of the Asian mainland. The herbivorous animals found include hippos, pigs, deer, tapirs, antelope, rhinos, and elephants (a species related to the Indian elephant as well as the stegodont elephants), as well as one of the last of the chalicotheres: unusual, massive, superficially horse-like beasts with large, well-developed claws on all four feet, found earlier in the Cenozoic throughout the northern continents and Africa. The range of meat-eaters was also broad: dogs, hyenas, (small) bears, tigers, and other cats—including sabercats and leopards—as well as civets, crocodiles, and monitors. Monkeys, apes, rabbits, various rodents, hawks, cranes, and turtles have also been found. None of the other islands has so far yielded nearly as large a fauna, but the remains of small elephants, giant rodents, snakes, tortoises, crocodiles, and monitors have been found. The faunae of the more eastern islands seem likely to have been less rich than that of Java because these islands were never connected to mainland Asia.

Simpson to the contrary, Wallace's lines do mean something. To the west live mammals like tigers, lions, and rhinos, and to the east, kangaroos, wallabies, and wombats. Some birds—weavers and barbets in the west, cockatoos and megapodes in the east—also recognize the boundaries. In the past, so did crocodylians, with mekosuchines in the east and *Crocodylus* in the west (at least up to the Early Pliocene, when the Indopacific croc probably arrived in Australia). Lizards, however, support Simpson's view, with agamids (discussed in chapter 4) and monitors living both east and west of the boundary.

Wallace's lines commemorate, if you like, the meeting of two lands with two separate florae and faunae: Asia to the west, Meganesia (Australasia) to the east. This chapter has attempted to indicate how different Meganesia was from Indonesia, a topic we shall pursue further in chapter 6. As these lands approached, the boundary between them, once represented by uncrossably vast distances of open ocean, became traversable. Rodents and lizards emigrated from one land to the other. As the continents continue to approach in the future, Wallace's lines will come to mean less and less as more and more creatures cross them. They are not a modern biogeographic feature, but a memorial of the past. And among the lizards that breached this boundary were the ancestors of *Megalania*.

# 3. The Discovery of *Megalania*

## History of Our Understanding of *Megalania*

### Initial Conceptions and Preconceptions

In keeping with the colonial status of nineteenth-century Australia, many fossils were sent back to England for study. In London, where a considerable mass of fossils from around the world had been accumulated at the British Museum, this permitted a more profound, less parochial understanding of the significance of Australian fossils in a global context. The realization that the world had only relatively recently been home to a variety of large and extraordinary animals (now extinct), often termed the *megafauna,* was in the process of being forged. Many Europeans had only just become aware of the variety of large animals still inhabiting the world, through the media of zoos and natural history museums. In regions outside Europe and North America, some of these animals—such as okapis and pygmy hippos—were still being discovered in the early twentieth century. In fact, in southeast Asia, other large animals are still coming to the attention of zoologists.

In the absence of any preconceptions, and with the knowledge that a megafauna had existed, scientists sought to understand the Australian fossil beasts. Initially—and plausibly enough at the time—some Australian animals were taken to represent creatures similar to those found elsewhere. Australia was clearly isolated now, but in an age when biogeographers had become aware of land bridges and considered the possibility of lost continents (like Lemuria in the Indian Ocean) on what then seemed good evidence, no one knew whether Australia had once been part of Asia or not.

As elsewhere, the first fossils from Australia often ranged from incomplete to very fragmentary. The doubly crested molars (grinding teeth) of *Diprotodon* resembled those already known from deinotheres and were initially believed to be from those beasts. Discovery of more complete remains soon dispelled this notion, and it became clear that the faunal input from Asia in recent (geological) times had been decidedly minor.

In the absence of our modern knowledge of continental drift, it was believed that the Australian fauna represented an ancient, possibly essentially Mesozoic Asian fauna that had survived by courtesy of the isolation. James Dana, a nineteenth-century American geologist, went so far as to describe Australia as a (surviving) "Triassic continent" (1895). This idea was consistent with the occurrence of relict forms, "living fossils," such as the Queensland lungfish and the bivalve *Trigonia*. Seemingly archaic creatures such as the platypus also supported this view. So it was no great surprise that fossils of giant reptiles appeared in recent sediments of the Darling Downs, an agricultural region 150 kilometers west of Brisbane in Queensland, Australia. They were first described as *Megalania prisca* (the ancient great strider) in the year of the publication of *The Origin of Species,* 1859.

The recognition that the Mesozoic was an age of reptiles came in the first half of the nineteenth century, but the realization that the dominant among these reptiles were dinosaurs came only with the great discoveries in the western United States after the American civil war. In fact, serious comprehension of the nature of dinosaurs and their relationships to other reptiles only began in the years just before the First World War. Thus, in the nineteenth century it was not seen as anomalous that giant reptiles had lived in Australia and that these creatures had nothing to do with dinosaurs. Furthermore, the prevalence of marsupials in Australia and their discovery in the Mesozoic beds in England led to the conception that they too were an archaic, Mesozoic kind of animal. Ironically, we now know that these Mesozoic English mammals were not marsupials.

## Illusion and Confusion

Sir Richard Owen—probably the preeminent anatomist of the nineteenth century (at least in English-speaking countries) and describer of *Megalania prisca*—was mindful of the relationship of extinct to living organisms, even though he was not a convert to Darwinian evolution. He compared some of the Australian reptilian remains to those of a small but baroquely spined agamid lizard of the outback desert, the so-called thorny "devil" (*Moloch horridus*). In 1886 he proposed a second "megalanian" reptile, *Meiolania*. The early remains of *Megalania* were distinctly sparse—in fact, they still are, as are those of *Meiolania*—so it was not until almost thirty years later that Sir Arthur Smith Woodward, a paleoichthyologist at the British Museum, showed that two different kinds of animals had been conflated. One was the giant monitor *Megalania,* and the other, *Meiolania,* was a giant tortoise.

*Moloch* had turned out to be a misleading subject for comparison, and the spined skull attributed to *Megalania* was actually from the tortoise. Owen had initially thought *Megalania* was related to *Hydrosaurus* (the latter now recognized as a species of monitor, *Varanus*), in other words that it was a monitor. But the discovery of what he believed to be the skull induced him to reinterpret *Megalania* as a giant simulacrum of the agamid *Moloch*. He was apparently slightly bemused by the absence of teeth in what he thought was *Megalania* since *Moloch* does have small teeth. Giant horned tortoises were of course unknown when he studied these fossils in about 1879, so it is perhaps understandable that he took the skull for that of the great lizard. However, George F. Bennett, who collected the skull, believed it to represent a giant turtle (cf. Gaffney 1983, which includes a history of the understanding of *Meiolania*).

Toward the end of the nineteenth century a similar skull was found in Cretaceous (or possibly Early Tertiary) deposits in Argentina and was interpreted as that of a turtle by Smith Woodward. Similar material had earlier been found in Late Pleistocene beachrock on Lord Howe Island off the eastern coast of New South Wales. This was studied by Thomas Henry Huxley, who in 1887 recognized it as belonging to *Meiolania* (even though he proposed to call it *Ceratochelys*). He also realized that this was essentially the same kind of animal, a tortoise, as that from which the skull that Owen had attributed to *Megalania* had come. Smith Woodward pointed out that even mammals were involved: Owen's last paper on *Megalania* included misidentified foot elements (probably) from a giant wombat.

Perhaps because Owen may have thought of Australia as a surviving province of the Mesozoic—a view that until recently was still useful for science fiction novels such as *The Ant Men* (1955) by the Australian Charles Cronin ("Eric North")—he did not find it surprising that more than one giant lizard might have lived there. So when in 1883 the fragment of a jaw was sent to him by Robert Etheridge Jr., a geologist in New South Wales, he took it to represent a second giant saurian, *Notiosaurus dentatus*. This he recognized as a giant monitor, as he had originally thought *Megalania* was. By 1888 Owen was aware that things were not as he had earlier supposed, and he proposed that *Megalania* and *Meiolania,* which he still thought to be closely related, were examples of a group of reptiles allied to both lizards and turtles. These he called "ceratosaurs," unaware that this name had earlier been applied by the American paleontologist Othniel Charles Marsh to a group of carnivorous dinosaurs (a use still current). But by this time the relationships of the various fossils were rather clear, and Owen's new suborder was never generally accepted. Owen had argued that the complete absence of a shell or of any evidence for a shell indicated that these remains did not pertain to a turtle. Eventually, however, the shell did turn up. But this came from Lord Howe Island, not the Australian continent, where remarkably little material from the shell has ever been found. The situation ultimately turned out to be even more complex, for the horned tortoise from the Darling Downs is *Ninjemys*, not the same as the *Meiolania* found on Lord Howe (although fossils of *Meiolania* have been found on the Australian continent in Queensland and the Northern Territory)

*Ninjemys*—its name indicating (the presumably original) "ninja turtle"—is, in its way, a vexing animal. Other than the original skull and tail club found by Bennett in the late nineteenth century, only a single vertebra of this kind of tortoise has ever been found again on the Darling Downs (from a different locality), and that despite much searching by professionals and amateurs alike. Against logical expectation, no remnants of a shell have ever appeared. Apparently, the animal was very rare, but that does not seem to explain the absence of what must have been durable pieces from a shell, when fragments from the shells of freshwater chelid turtles are not uncommon.

This much of the history of the understanding of *Megalania* is reasonably well known, at least among paleontologists who work on lizards, snakes, and their relatives (lepidosaurs). Paleontologists in Australia as well as in England were involved, particularly Etheridge in New South Wales and Charles Walter de Vis in Queensland. Both of these gentlemen were immigrants from England. The latter, who had apparently "upgraded" his name to "de Vis" from Devis (or even Davis) upon arriving in the colony, could be an inspiration for the aging population of our time. Almost all of his scientific work was done after his fiftieth birthday in 1879. If he was exemplary in that, in the late nineteenth century he was also considered a somewhat bumbling colonial "expert," who lacked the expertise of those in England. Indeed, he did, but in view of the errors made even there and the resulting misinterpretations—and, not least, the lack of resources in Australia—he seems to have been more competent than he was generally given credit for, at least in paleontology.

Like Huxley, and at almost the same time but apparently independently, de Vis realized that Owen's *Megalania* was a chimera, an inadvertently fanciful beast. In 1889 he published his reasons, which were rather commonsensical. All the bones known looked like those of monitors except the skull; thus, the skull did not pertain to *Megalania.* De Vis noted that if *Megalania* was truly a giant simulacrum of *Moloch,* one would expect that it, like *Moloch,* possessed spines in its skin. Given the size of *Megalania,* these should have been ossified, like its cranial horns. (The "horns" and dermal spines in *Moloch* are not ossified, but keratinous.) Such dermal spines have never been found, presumably because they never existed.

Some years earlier, in 1885, de Vis had mentioned a giant monitor that he called *Palvaranus brachialis,* at a meeting of the Royal Society of Queensland. The contents of this meeting were never formally published by de Vis, but a report appeared in the Brisbane newspaper on 14 March 1885. From this report it is evident that at this time de Vis also looked upon Australia as having, until recently, remained in the age of reptiles. He regarded *Megalania* as a giant agamid and proposed two giant varanids: Owen's *Notiosaurus,* and his *Palvaranus.* The latter was based on bones of the forelimb, and so not comparable to the jaw that represented *Notiosaurus.* On reflection, however, he concluded that there was a superfluity of large reptiles here, as he recorded in 1889, and he never formally published the name *Palvaranus.* At the time of the 1885 meeting—as he later explained—he had been under the misapprehension that the deposits of the eastern Darling Downs were Pliocene and thus older than those that yielded *Notiosaurus.* So later in the year 1885, he referred the remains of '*Palvaranus*' to

*Notiosaurus.* In 1889, de Vis went on to point out that the jaw of *Notiosaurus* "fulfils his Cuvierian anticipation 'that on the very probable hypothesis that the jaws and teeth of *Megalania* are of the same type as those of *Hydrosaurus,* it must have been carnivorous.'" Cuvier, as you doubtless remember, had proposed that animals being integrated entities, it was possible to deduce the nature of the whole creature from a part. (This is probably true, but not in as straightforward a fashion as Cuvier believed.) Hence, de Vis's turn of phrase in hypothesizing the diet of *Megalania.* Thus, *Notiosaurus* too was *Megalania,* a judgment that has been universally accepted since. De Vis did, however, propose that a tooth from a giant monitor represented a new species, *Varanus dirus.*

To summarize this convoluted history, let's track the basic concepts involved. Owen first, in 1859, conceived *Megalania* as a giant monitor, a lizard. With the discovery of the horned skull, he then, in 1880, reassessed it as a giant agamid, again a lizard. He resurrected the concept of the giant monitor in 1884(b), as *Notiosaurus.* De Vis conceived a second giant monitor, *Palvaranus.* New discoveries enabled Owen to propose *Meiolania* in 1886. The following year Huxley realized that the horned skull from Queensland was that of a tortoise. And in 1888 Woodward chronicled the confusion and set the record straight: a brief version was later (1889) published by Richard Lydekker at the British Museum. Also in 1888, Owen proposed that *Megalania* and *Meiolania* represented a previously unknown group of giant reptiles, the ceratosaurs. In the end, the giant agamid—and the ceratosaurs (in this sense)—proved nonexistent, but both the giant monitor (one of them, anyway) and giant horned tortoise were real.

In 1889, however, de Vis proposed another giant monitor, *Varanus dirus.* And in 1899, A. Zietz, at the South Australian Museum in Adelaide, proposed yet another giant monitor, *Varanus warburtonensis.* Zietz thought this was not *Megalania* because of its smaller size, since the single vertebra known was only about two-thirds as large as one of *Megalania* from the Darling Downs. Our current understanding of what belongs to whom is given in Table 3.

The name '*Palvaranus*' indicates "ancient monitor," and '*Notiosaurus*' indicates "saurian of the south," both quite appropriate. But it was never clear why Owen chose the name *Megalania* ("great runner") or, for that matter, *Meiolania* ("moderate runner"), for creatures that, as far as we know now, no one ever thought to be particularly swift.

### Early-Twentieth-Century Studies

The Hungarian paleontologist Dr. Baron de Fejervary G. J. in 1918 claimed that Owen recognized, in his 1888 paper, that *Meiolania* was a turtle. (In Hungarian, as in Chinese, the given names follow the family name, thus we refer to him as de Fejervary G. J., not as G. J. de Fejervardy.) Owen wrote no such thing in this paper, and it is belied by his 1888 paper on *Meiolania,* proposing the ceratosaurs, which— although related—were clearly not turtles. So de Fejervary must have been mistaken.

It is perhaps mildly ironic that, although Owen was incorrect in his

## Table 3
### The currently accepted identifications of material attributed to the Australian giant monitor

| As originally described | As now accepted |
| --- | --- |
| *Megalania prisca*, postcranial bones (except some foot bones) | *Megalania prisca* (varanid) |
| *Megalania prisca*, foot bones | *?Phascolonus* sp. (giant wombat) |
| *Megalania prisca*, skull | *Ninjemys oweni* (chelonian) |
| *Notiosaurus dentatus* | *Megalania prisca* (varanid) |
| '*Palvaranus brachialis*' | *Megalania prisca* (varanid) |
| *Varanus dirus*, holotype tooth | *Megalania prisca* (varanid) |
| *Varanus dirus*, referred maxilla | unknown (varanid) |
| *Varanus emeritus* | *Varanus* (varanid), possibly *V. salvadorii* |
| *Varanus warburtonensis* | *Megalania prisca* (varanid) |
| *Meiolania oweni* (Qld.) | *Ninjemys oweni* (chelonian) |
| *Meiolania platyceps* (Lord Howe Is.) | *Meiolania platyceps* (chelonian) |

interpretation of several giant lizards (or even stranger reptiles, the "ceratosaurs") living in Pleistocene Australia, several giant lizards did—and one still does—live in the southwestern Pacific region. Some twenty-three years later, a living giant monitor, the ora (*Varanus komodoensis*) was found in Indonesia. Found by Europeans, that is—the Indonesians had known of it for a long time, and perhaps it has even influenced our concepts of mythological dragons. Moreover, in the past few years fossils of a recently extinct giant iguana (about which more later) have been found in Fiji.

The next appearance of *Megalania* in the scientific literature after 1889 was by Etheridge in 1917. By this time the issue of what *Megalania* was had long been settled, and if the issues that were raised were not entirely new, they were at least different. Etheridge described a new vertebra that turned up in the deposits in Wellington Caves in the hills west of Sydney in New South Wales. This was the first report of megalanian fossils in cave sediments; previously they had been found in stream or riverine deposits. Almost eighty years later, in the late spring (December) of 1995, a small symposium was held at the caves—the opening session actually held in one of them—and another vertebra of *Megalania* was found in the cave earth (but in a different section of the cave system and after the session). This caused some excitement, since none of us present remembered Etheridge's paper.

Etheridge regarded *Notiosaurus* as a synonym of *Megalania* and also raised again an issue hinted at by de Vis and previously mentioned by Lydekker, namely the validity of *Megalania* as a genus. Was *Megalania* valid, or was it actually *Varanus priscus*? In other words, was *Megalania* sufficiently different from the species of *Varanus* that it belonged to a distinct genus? Lydekker, without comment, had in 1888 simply classified *Megalania* as *Varanus priscus,* apparently because of its similarity to *Varanus sivalensis,* from the Pliocene of northern India. This is a point not yet finally resolved, but since the rise of phylogenetic systematics (cladistics) has led us to look at it in a new way, we shall return to this issue.

In 1918, *Megalania* was again chronicled in Europe, by de Fejervary. He wrote two monographs on fossil varanids, the other appearing in 1935. De Fejervary concluded from a study of the vertebrae that not only was *Megalania* a valid genus (and not a species of *Varanus*), but that it did not even belong to the same family. He argued that the presence of accessory articulations on the vertebrae (zygosphene and zygantrum) indicated that *Megalania* was quite different from varanids, and he also considered the relatively small size of the neural canal a significant difference.

A single vertebra may be looked upon as consisting of two (conjoined) portions. The lower, spool-like body (or centrum) usually supports the weight of the animal. Above this is an arch of bone that surrounds the spinal cord, the neural arch, to which is attached a suite of three large and four small projections. Two of the projections, the transverse processes, extend symmetrically to the sides to contact the ribs and their associated muscles. The other, the neural spine, extends upward, and serves as an attachment point for the muscles of the back. The four small projections (called zygapophyses by anatomists) are found in two pairs, one pair at the front (called prezygapophyses) and one at the back (postzygapophyses). Those at the front of a vertebra contact those from the back of the preceding vertebra and help limit the movement in the vertebral column. The bodies of any two adjacent vertebrae are also in contact, joined by the pulpy intervertebral discs (a ruptured intervertebral disc is known as a "slipped disc").

This is the normal condition for animals, including people, but some creatures, especially snakes and some large dinosaurs, develop other, accessory contacts between the vertebrae. These, associated with the zygapophyses, are called the zygosphene and zygantrum. Since these presumably enhance the strength of the vertebral column, appropriate in a very big monitor, why they should indicate that *Megalania* wasn't a giant *Varanus* is not entirely clear. However, it is not obvious that *Megalania* actually had zygosphenes and zygantra; in fact, there seems to have been some misunderstanding here. Zygosphenes and zygantra are accessory joints in which a wedge of bone on the front of one vertebra (the zygosphene) contacts a recess (the zygantrum) on the preceding vertebra. Such structures do not exist in *Megalania* at all—at least not on any of the Queensland Museum specimens. If anything, it seems the wedge would have been on the back of the vertebra (and hence properly called a hyposphene) and the contact for it (the hypantrum) on the front of the succeeding vertebra. Sometimes what seems to be a hypantrum is clearly present, but so far as I have seen, never a hyposphene. Even a good example of a hypantrum is rare.

De Fejervary regarded *Notiosaurus* as *Megalania*. Incidentally, in 1918 he published the first historical remarks (since those of Woodward) about the discovery and subsequent progress in understanding *Megalania*. He also followed Owen in attempting to understand the paleobiology of the beast, specifically its large size and extinction. He was impressed by fellow Hungarian Baron Nopcsa Ferenc's recognition that large dinosaurs often had disproportionately large pituitary glands (hypophyses). In the human pathological condition of acromegaly, caused by a malfunction of the pituitary gland, certain parts of the body grow disproportionately, and this can result in very large individuals (appropriately called acromegalic giants). Nopcsa thought the dinosaurs—and de Fejervary that *Megalania*—grew so large because of an inherited dysfunction of the pituitary.

De Fejervary foreshadowed the philosophy of science propounded by Karl Popper, popular during the 1970s and 80s and—rightly—still influential today. One of Popper's ideas was that scientific hypotheses should be stated in such a way that any evidence potentially contradicting them becomes clear. In other words, the kinds of observations that would show hypotheses to be wrong should be clearly set out. If no such evidence can be found, we may assume that the hypotheses are correct (until such evidence shows up). The good Baron suggested that if his and Nopcsa's notions were correct, it would be expected that *Megalania* had a proportionately smaller head compared to modern varanids and that the phalanges (toe bones) would be relatively thicker. He thought that the head was small but was unaware that any phalanges had been found. In fact, the phalanges do not seem unusually thick. Since the cranial bones—with a single unhelpful exception—cannot be associated with postcranial remains, we can not tell whether the head was disproportionately small, since we do not know the size of the body from which the cranial bones derived. Since acromegaly also involves a decrease in sexuality, de Fejervary thought that this might also explain the extinction of this animal. We now might well think this quaint, but we also have a better understanding of the time scale of the earth's history. In 1918 the dinosaurs were thought to have become extinct about 20 million years ago, but we now date this at 65 million years ago. The old, abbreviated history makes more plausible Nopcsa's notion that the lizards inherited a pituitary dysfunction that on the one hand resulted in their gigantism, and, on the other, resulted in a slow decrease in reproduction that eventually led to their extinction. It is not nearly as much a burden on the imagination to suppose that the lizards may have slowly died out over a period of 500,000 years from acromegaly than that they survived with it for 4,000,000 years, almost ten times as long.

De Fejervary died in 1932, but he had continued to work on varanids and *Megalania* after 1918, and the results of his researches appeared posthumously. Regarding *Megalania,* his second monograph (1935) presents a detailed description of a vertebra and reiterates his earlier conclusion that it represented a distinct family, although he does recognize its similarity to *Varanus sivalensis.* Among other issues covered is the description of two vertebrae from the Kendeng beds of Java as *Varanus bolkayi,* the first indication of fossil monitors from Indonesia. De Fejervary also regarded the ora as a genus separate from *Varanus* and named it *Placovaranus.* He did so because he believed that its osteoderms, small bony nubbins embedded in the skin, were not found in *Varanus.*

### Modern Understanding

The first modern (post–World War II) study was carried out by the jaunty New Yorker Max Hecht and published in 1975. Hecht pointed out features that indicated that *Megalania* was a varanid as well as those that he thought distinguished it from *Varanus.* Most of the latter seem to be related to the large size of *Megalania.* Hecht described all of the material then available in Australian museums, mostly in the Queensland Museum. Like de Fejervary and Owen, he attempted to understand the creature's ecology. Unlike them (they had insufficient material to do so), he also tried to reconstruct its evolutionary history. Hecht pointed out that fossils of *Megalania* were often associated with those of large mammals and that it was—like the ora—probably a predator of them. He was perhaps the first to appreciate that having a lower metabolic rate during cool periods allowed these huge lizards to require less food than mammals of similar size—a feature we shall discuss later. Of course, this would often be true during warm periods, as well. Their evolutionary history was, he thought, one of taking the opportunities to exploit available untapped resources (particularly sources of food), rather than being constrained by some inherent "physiological and behavioural limitations" of being a lizard, as had often previously been thought. In other words, he interpreted it in ecological terms rather than in terms of intrinsic capabilities (or their lack).

Just before the study of Hecht appeared in 1975, Tom Rich, an expatriate Californian paleontologist at the Museum of Victoria in Melbourne, undertook to reconstruct the complete skeleton of the lizard (Fig. 3.1). Rich described this endeavor in a joint article with Brent Hall, a preparator at the museum, published in the magazine of the Australian Museum, *Australian Natural History,* in 1979. They chose to reconstruct a skeleton 5.7 meters (ca. 18 feet 8 inches) long. (Their article says that the skeleton was 5.5 meters long, but the one on display in the Queensland Museum is 5.7 meters.) This provided for the first time a graphic representation of what *Megalania,* or at least its skeleton, looked like, and it drove home the immense size of the beast. Tom's reconstruction, more than Hecht's paper, is probably responsible for our current image of the animal.

*Fig. 3.1. The restored skeleton of* Megalania prisca.

## History of the Discovery of *Megalania*

The history of the discovery of *Megalania* parallels the history of its understanding—not surprisingly. Most specimens have been found by interested rural folk or by professional or amateur collectors. Prior to 1889, probably the year before, Mr. R. W. Frost donated to the Queensland Museum a small series of bones that he had collected. Frost lived on the eastern Darling Downs near Kings Creek and reported to de Vis that he had found the specimens all together. Just a few years later, in 1892, Herman (or Hermann?) Lau, who had come to Australia from Germany and was returning home, presented the Australian Museum with a set of bones that he had collected in 1882. The fossils probably came from the bank of Kings Creek (his letter implies but does not explicitly state this) near Pilton. In examining these specimens, Hecht noted that the Frost material included a left ulna, and the Lau collection a right. Hecht felt these two bones were very similar in size and proportions and thus might have derived from the same individual: no elements were duplicated in the two collections. These were the only specimens that probably represented partial skeletons (or a single skeleton) that Hecht could find. Many of the specimens of *Megalania*, like those of Frost and Lau, were found on the Darling Downs. Although fossils of *Megalania* were found over much of eastern and central Australia, these tended to consist of just one or a few bones.

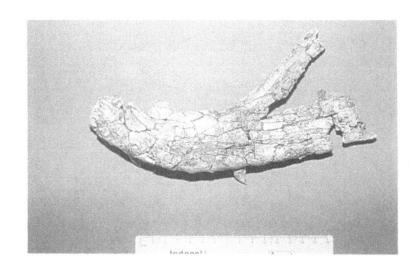

Fig. 3.2. *(top) The left maxilla of* Megalania, *seen in dorsolateral view. This specimen, found by Henk Godthelp on the Darling Downs of southeastern Queensland, is the most complete maxilla of* Megalania *yet discovered.*

Fig. 3.3. *(center) Two left frontals of* Megalania, *seen from below. These are the first discovered, found by Ian Sobbe, also on the Darling Downs.*

Fig. 3.4. *(bottom) The two left frontals of Fig. 3.3 seen in medial view. The lower, smaller specimen was much worn before discovery. The marked convexity of the upper is the sagittal crest.*

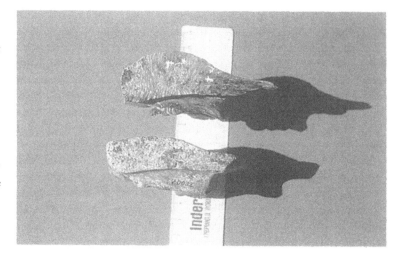

The next step forward in understanding *Megalania* came with the discovery of more, and new, fossils. Recent discoveries, like those of the nineteenth century, were made by nonprofessionals. Henk Godthelp, a tall, thin high school science teacher in Brisbane before joining the staff of the University of New South Wales in Sydney (and now at the Australian Museum), discovered in the early 1980s the first significant new cranial remains found since the nineteenth century. This was an almost complete maxilla, the major bone of the upper jaw, found on the Darling Downs (Fig. 3.2). Ian Sobbe, a farmer on the Downs whose property borders Kings Creek, is an inveterate and enthusiastic amateur paleontologist and collector of fossils. By the late 1980s he, like many other paleontologists (especially in Australia), had accumulated a fair number of unusual specimens that he was unable to identify. In 1988 he made the day trip to Brisbane to see if some of the specimens could be identified. Among them were two bones of the skull roof of a giant varanid, the first frontals ever discovered of *Megalania* (Figs. 3.3, 3.4). Ian returned west to his home on the Downs and, searching through the remainder of his unidentified specimens, turned up a parietal of *Megalania*. Together with a braincase discovered in the nineteenth century and held by the British Museum (Natural History), this allowed a much more complete understanding of the skull of *Megalania* than was possible before.

But the latest episode in the history of the discovery of this unusual lizard was, in its way, more bizarre than any that preceded it. It happened in the mid-1990s with the arrival of Mr. J. Hawkins, a geologist, at the museum in Brisbane. Some thirty years before, in 1965, French petroleum geologists prospecting for oil in southeastern Queensland had found three dorsal vertebrae—those from the back—of *Megalania* at Alderbaran Creek, near Springsure in eastern Queensland. They had duly reported the discovery to the Queensland Museum and donated the specimens. The museum paleontologist then visited the creek, found the site, and collected the remaining material: a few pieces of rib, a few isolated teeth, and a small piece of maxilla. The next day, the 12-year-old son of the manager of the land on which the fossils had been found went out to the site and conducted his own excavation. He uncovered at least 10 ribs, a broken pelvic girdle that is still the most complete known (Fig. 3.5), a femur (Figs. 3.6, 3.7), more of the maxilla, five incomplete and unidentified bones, and more than fifty small pieces of bone. The maxillary pieces he found fitted to that collected by the museum paleontologist. The 12-year-old grew up, studied at the university, and became a geologist—and then, remembering that he still had the material he had collected, donated it to the museum.

This is arguably the most informative specimen of *Megalania* ever found, and all owing to the interest of a 12-year-old. It seems to represent the incomplete skeleton of a single individual, the third or maybe even the second (if the Frost and Lau specimens really are from the same individual) ever found. The reason it is the most informative will be made clear in part in the following chapters. But the discovery of this specimen also afforded enough additional bones to permit some to be prepared for microscopic (histological) examination. This re-

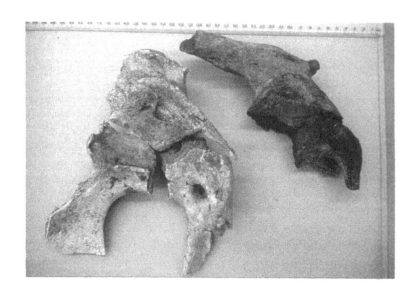

Fig. 3.5. (top) The two pelves of Megalania. That on the right was found in the late nineteenth century on the Darling Downs and has an almost complete ilium and part of the pubis, but the ischium is missing. That on the left, found by J. Hawkins at Alderbaran Creek, has the body of the ilium and most of the ischium and pubis. Both specimens were found in eastern Queensland, and both were remarkably recent in appearance.

Fig. 3.6. (bottom) The femur of the Alderbaran Creek Megalania.

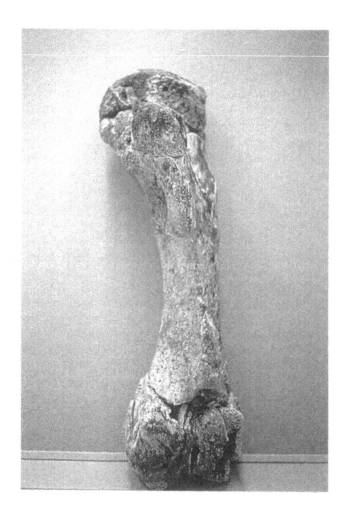

Fig. 3.7. *The proximal end of the femur (right) of the Alderbaran Creek* Megalania, *with the proximal epiphysis removed (left). That the epiphysis was not fused to the shaft of the femur may indicate that this was an immature individual, but in some modern monitors, the epiphysis never completely fuses to the shaft.*

search, still underway, should provide information on the life history and age of this individual.

## Biogeography of *Megalania*

*Megalania* has been found throughout eastern Australia as far west as the bed of Lake Eyre, as far south as the Melbourne area near the southern coast, and north into the Cape York Peninsula of Queensland. It has not been reported from Tasmania or New Guinea. The land bridges linking those islands with the Australian mainland are thought to have sustained habitats unlike those of the surrounding higher country, the country which, with the rise in sea level, became the modern islands and mainland. As mentioned in the previous chapter, these habitats are thought to have acted as barriers since, at least in the north, many animals that seemingly could have moved from New Guinea to Australia and vice versa in fact did not. Perhaps *Megalania* was unable to cross these barriers, or maybe it did live in New Guinea or Tasmania but its fossils have not been found (or were never preserved). If it got as far west as Lake Eyre, there seems no obvious reason why it did not live in Western Australia, but so far there are no fossils.

*Megalania* seems to have been rarer in the south than in the north. There are only three localities in New South Wales and Victoria, and there are more sites yielding *Megalania* in Queensland than in all the other states put together. If we take this at face value—and there are good reasons why we shouldn't—it suggests that *Megalania* was concentrated in the northeastern part of Australia. However, it is clear that it also lived in the episodically (and presently) arid center of the continent.

The deposits yielding fossils of *Megalania* were mostly laid down by streams or rivers. But some fossils have been found in cave deposits. Fossils of *Megalania* are only rarely found in either situation, which implies that the creature did not habitually frequent watercourses or caves. In contrast, fossils of the crocodilian *Pallimnarchus* have been found in riverine deposits but never in cave deposits, suggesting that *Pallimnarchus* lived in or near rivers. The occurrence of its fossils is consistent with *Megalania* having ranged broadly over the countryside.

## History of Lizards in Australia

The largest lizard in the world (after the Mesozoic mosasaurs) was still a lizard. And Australia is a place where lizards form—and presumably always formed—an important component of the fauna. Thus, the history of lizards in Australia may provide clues about the occurrence and evolution of the largest lizard. Unfortunately, as has been said, the fossil record in Australia is not all that could be desired. Only in the past decade has there been any substantial improvement in our understanding of the history and evolution of reptiles in Australia, and this largely for crocodilians. Prior to this, much of the understanding was based on attempts to reconstruct evolutionary history from the patterns of speciation and biogeography of modern forms. When these inferences could be checked against fossils of some lineage, they proved to underestimate how long that lineage had lived in Australia.

The quality of the fossil record in general is inversely proportional to its age. For Cenozoic Australia, the Pleistocene record for lizards is the best and that of the Oligo-Miocene beds the poorest (and the earlier Cenozoic record is effectively nonexistent). Only two bones make up the pre-Miocene record of Australian terrestrial lizards, but a few mosasaur bones have been found in Western Australia. A single incomplete humerus, probably from a terrestrial lizard, was found in the Early Cretaceous rocks of Victoria (Fig. 3.8), and a skink femur from the Eocene of Queensland. The Cretaceous lizard was a relatively large beast, perhaps a meter long, but the articular ends of the bone are missing, so it can not be identified more precisely than simply as a lizard. These bones demonstrate that lizards were present in Australia throughout the Cenozoic and probably well before. They do not show whether the lizards now living in Australia evolved there from forms like those that left these two pre-Miocene fossils or immigrated from Asia.

In the absence of an adequate fossil record for the Cenozoic prior to the Miocene, most lizard lineages (as well as most other tetrapod lineages) in Australia appear to begin at that time. In a few cases, particularly for crocodilians and turtles, we do have some earlier record that permits a glimpse of their previous history. Unfortunately, all of the interesting lizard history (like much of that of marsupials) seems to have occurred prior to the Miocene, during the Paleogene. All of the Australian lizard families (agamids, geckonids, and scincids as well as varanids) are already represented in Miocene deposits. Even some living genera, such as the water "dragon" *Physignathus* among agamids, various skinks, like *Egernia, Eulamprus,* and *Tiliqua,* and—of course—the monitor, *Varanus,* are already present. The geckos remain unstudied.

Only three extinct species of Cenozoic lizard—a scincid belonging the *Sphenomorphus* group, the agamid *Sulcatidens,* and a pygopod (a limbless lizard related to geckos)—are so far known from Australia. It

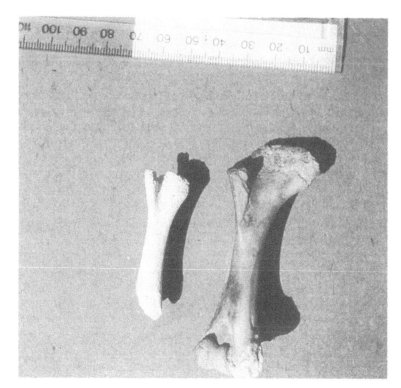

Fig. 3.8. A plaster cast of the humerus (incomplete) of a lizard from the Early Cretaceous rocks of Victoria (left), compared with that of a modern monitor (right). This incomplete humerus is the oldest record of a lizard from Australia.

would not be surprising to find other extinct species among the yet-unstudied fossils. The modern fauna of Australian lizards seems to be conservative and to include many forms that predate Pleistocene and perhaps even Pliocene times: a fauna, perhaps, of small, creeping living fossils.

## The Other Reptiles in Australia

The record of other Cenozoic reptiles in Australia is not substantially better than that of lizards. Among the non-lepidosaurians there is a good record of chelids (shells, at least) into the Eocene. *Niolamia*, the horned tortoise from the Cretaceous of Argentina, indicates that the meiolaniid turtle lineage in Australia must have substantially predated the (until recently) oldest-known fossils from the Pliocene. There has been no land connection between what is now Argentina and Australia since the Cretaceous, so meiolaniids must already have been in Australia somewhere. The discovery of the meiolaniid *Warkalania* in Oligocene or Miocene deposits at Riversleigh in northwestern Queensland confirmed this deduction. The already-mentioned indigenous radiation of crocodilians, the mekosuchine crocodilians, in Australia and other southwestern Pacific lands, produced many unusual forms. These animals became extinct during the Pleistocene in Australia but survived until the time of the Roman Empire in New Caledonia and almost as recently in Vanuatu.

Fossil snakes are not well known in Australia, but snakes are not the most common of fossils anywhere. However, a few large snakes are known from Australia, many representing groups still surviving, such as the pythons or the venomous elapids. One, however, was among the surprises of Australia's past. A group of boid-like snakes, the mad-tsoiids, had long been known from the Late Cretaceous and Early Tertiary of South America, Africa, and Madagascar. In Australia these snakes survived until the Pleistocene in the form of *Wonambi*. As far as snakes go, *Wonambi* was not a particularly large snake, but it has recently been shown to be one of the most primitive, a snake that would not be out of place among the dinosaurs of the Late Cretaceous. In summary, various Australian snakes (elapids, typhlopids, pythonids, and madtsoiids) date back at least into the Miocene, and some probably into the Cretaceous.

Much of the Tertiary reptilian fauna of Australia remains to be discovered. The Pleistocene fauna is better known, but in a land where even the modern lizard fauna still yields surprises—like the approximately 20-cm-long skink *Nangura spinosa* discovered in 1992 in what had been thought to be a well-explored region not far from Brisbane, Queensland's largest city (Covacevich, Couper, and James 1993)—we cannot pretend that it is adequately known. Nonetheless, in order to see what was happening in evolutionary terms rather than just list when various forms appeared, we can analyze the record.

The Pleistocene fauna can easily be divided into two components. There are several ways to do this, but the obvious criterion is extinction and survival: dividing the fauna into the lineages that still survive today and those that do not. We may also divide them by size, into those that were substantially larger than their living relatives—creatures like *Megalania*, *Wonambi*, *Meiolania*, and *Ninjemys*—and those that are not (recognizing that we are stretching the truth a bit in the case of the snake *Wonambi* since the living Amethystine Python—*Morelia amethystina*—grows equally long). Unexpectedly, the resulting groups in both cases are almost the same: in other words, the "giants" became extinct.

We must be cautious about this, however, because the fossils are poorly known and the small forms are often represented only by fragments. Thus, extinct small forms may have gone unrecognized or undiscovered. One large "Pleistocene" reptile—but only one—did survive: the Indopacific crocodile *Crocodylus porosus*. The trionychids (soft-shelled turtles) also may not conform. They had been in Australia at least since Eocene times, although there is no sign of them (yet) among Cretaceous fossils. Their origin in Australia is unclear: they seem not to have lived on the Gondwanan continents during the Mesozoic. Perhaps they are unexpectedly early Asian "immigrants." Although large, they were not large enough to be considered giants but still possibly large enough for their size to account for their extinction. The problem is that we don't really know why the giant forms became extinct. Still, this analysis provides a starting point for our understanding.

There is yet another set of criteria that may be used: when the lineage originated. There are three possible groups here: first, those that originated in, or migrated into, Australia before the breakup of Gond-wanaland—what we may call "Gondwanan relicts." This component consists of relict forms that had relatively low rates of evolution and would be the most closely related to the old Gondwanan (now South American or African) forms. These would date back before the final break between Australia and Antarctica about 55 million years ago. Second are those that originated in Australia between the time of its separation from Antarctica and the opening of a route from Asia—we can term these "Gondwanan derivatives." This component would be related to South American or African forms but have undergone evolutionary change. The second component would have had higher rates of evolutionary change, and its lineages would be between about 60 and about 5 million years old. Third are the immigrants from Asia, probably less than about 5 million years old in Australia—unless, of course, like some reptiles, they were adept at dispersing long distances along island chains.

Dividing the reptiles into three groups based on their time of origin also produces groups that are curiously similar to those generated by the two other sets of criteria. Of the giant forms, most—*Wonambi, Meiolania, Ninjemys,* and *Pallimnarchus*—seemingly are Gondwanan relicts (the first three) or derivatives (*Pallimnarchus*). *Megalania* and *Crocodylus porosus* are not, both lineages having arrived from Asia. Unfortunately, applying the criterion by origin requires a more complete fossil record than often exists. Thus, with the exception of the meiolaniid turtles, madtsoiid snakes, and mekosuchine crocs, we cannot distinguish Gondwanan relicts from Gondwanan derivatives. Even so, it appears that the surviving Australian lizards include no Gondwanan relicts: the Gondwanan relicts and giant forms tended to become extinct during the Pleistocene. The survivors are (some) Gondwanan derivatives and Asian immigrants that were not giants (except for the Indopacific croc). Why this is so, when at least some Gondwanan relict lineages survived successfully from at least 50 million to at least 50,000 years ago, is still a mystery.

# 4. *Megalania* and Other Varanids

## The Place of Monitors Among Lizards

A giant lizard is still a lizard, and we are all familiar with lizards. But we are familiar with living lizards, and this "familiarity" belies the fact that lizards, in the evolutionary sense of from whom they evolved and when, have only come to be understood in the past twenty years or so. Lizards, together with snakes, form a group of animals known as *squamates.* The term indicates "scaled ones," in reference to their (usually) scaly skins—an obvious feature. Squamates are the only surviving lepidosauromorphs, one of the four lineages (Fig. 4.1) that derived from the ancient tetrapod lineage that evolved the amniote egg (the egg with a set of internal membranes that could be laid and hatched on dry land). The other three lines include the archosauromorphs, now represented only by birds and crocodilians but including the dinosaurs and other obscure but unusual kin; the synapsids, now represented by the mammals but also including their extinct ancestors and related animals; and the turtles (and their extinct ancestors). Lepidosauromorphs go back at least 250 million years (to Permian times), but lizards are rather younger. They first appear during the Jurassic, between 150 and 200 million years ago.

For most of the twentieth century it was generally thought that lizards dated back to the Permian, for an interesting reason. The lizard-like body form—small to moderate size with long body, short neck, more-or-less short sprawling limbs, and long tail—is much older than lizards themselves. This form dates back into the Carboniferous, over 300 million years ago, and has been successively adopted by different groups of tetrapods, including some amphibians, for example, the salamanders. The lizards now seem to have the option on this form (except for the salamanders), and they have had it for longer than any of their predecessors (except, maybe, for the salamanders). So although

Fig. 4.1. (top) The four lineages of amniote tetrapods. From left to right these are: archosauromorphs (represented by a bird); lepidosauromorphs (represented by a lizard); testudinates (represented by a terrapin); and synapsids, mammals and their ancestors and kin (represented by a wombat). This arrangement is generally accepted, but some feel that the positions of synapsids and testudinates should be interchanged—that turtles and their kin diverged from the common lineage before mammals and their kin. The non-amniote tetrapod looks on.

Fig. 4.2. (center) Two lizard-like forms. Thadeosaurus (left) is from the Permian of Madagascar. This animal was about two-thirds of a meter long. The paliguanid, seen from above, is based on a composite skeleton in Carroll (1988). This was a much smaller creature, maybe 0.2 meters long. The distal tail skeleton is unknown, hence the docked tail.

Fig. 4.3. (bottom) Phylogenetic relationships among the (living) varanoid lizards. Varanids (V) are most closely related to Lanthanotus (L), and these together are related to the helodermatids (H). This diagram is based on the work of Pregill, Gauthier, and Greene (1986).

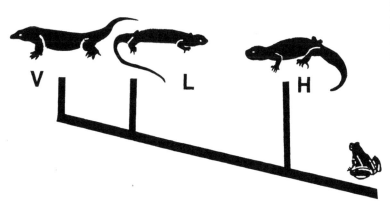

we may be familiar with living lizards now, if we could journey back in time to the Triassic and Permian, we would also recognize animals that looked and lived like lizards—but were not lizards (Fig. 4.2).

The relationships of the lizards known as monitors (the varanids) have long been clear in general but rather obscure in detail. In 1986 Gregory Pregill of the San Diego Natural History Museum—together with colleagues Jacques Gauthier and Harry Greene—found that *Varanus* was most closely related to the beaded lizards (the helodermatids, which include the Gila "monster"). They focused on helodermatids rather than monitors, so although their results are important to understanding monitors, they did not include snakes or other "nonstandard" lizards in their analysis. They found that, as was generally thought, the so-called "earless monitor" (*Lanthanotus borneensis*) of Kalimantan, in Indonesia, was the most closely related (the sister group) to the living monitors (Fig. 4.3). The beaded lizards were the next most closely related animals. An extinct group of monitor-like lizards, the necrosaurids, were the closest relatives of the lineage of varanids and helodermatids.

Two years later a book edited by Richard Estes and Pregill presented a massive phylogenetic analysis of living lizards by Estes, Kevin de Queiroz, and Gauthier. Their preferred "conservative evaluation" of lizard evolution recognizes four major lineages of lizards (Fig. 4.4): (1) iguanians (including chameleons and agamids as well as iguanids); (2) geckoes (including their limbless relatives, the pygopodids); (3) scincoids plus lacertoids (the skinks and the European animals to which the name "lizard" was first applied); and (4) anguimorphs. The latter lineage includes the monitors in addition to anguids and xenosaurs. This arrangement was not surprising, since Californian Charles Camp had expressed similar relationships in his classification of 1923. Estes and company, however, had some difficulty with their analysis when snakes (and the limbless amphisbaenians and dibamids) were included.

Anguids are not particularly obscure animals, but amphisbaenians, dibamids (Fig. 4.5), and xenosaurs are little-known reptiles, even to many herpetologists. Anguids include the alligator lizards (*Gerrhonotus*) and galliwasps (*Diploglossus*) of the New World: slender, sleek lizards with relatively small limbs. The Old World anguids have lost their limbs entirely: these are the so-called glass snakes, *Ophisaurus*. Xenosaurs (*Xenosaurus*) are robust Central American lizards, up to 38 cm (about 15 inches) long. Amphisbaenians, the "worm lizards," are also limbless like the dibamids, and are also not snakes. They are mostly found in North and South America and Africa, with but a few species in southern Europe and Indochina. Dibamids from Indochina, the Malay Archipelago, and New Guinea are burrowing forms that actually are not quite limbless; the males have flap-like hindlimbs. The xenosaurs and anguids are related to monitors, but the amphisbaenians and dibamids appear not to be.

Michael S. Y. Lee, then at the University of Sydney, most recently (1997) studied the relationships of varanids in conjunction with his conclusion that the closest relatives of snakes were not the monitors, as usually thought, but the mosasaurs (Fig. 4.6). These were a group of

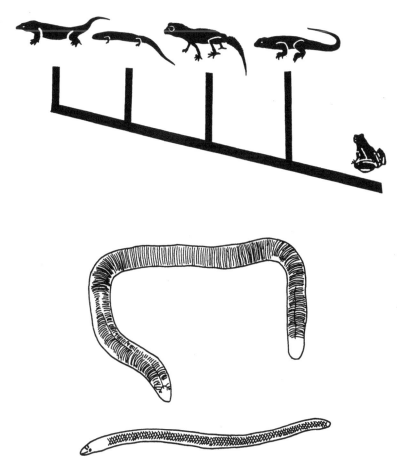

Fig. 4.4. (top) Relationships of the four major lineages of lizards, based on the work of Estes, de Queiroz, and Gauthier (1988). From left (most derived) to right, these are anguimorphs (represented by a monitor), scincoids and lacertoids (represented by a skink), geckoes (represented by a gecko), and iguanians (represented by an iguana).

Fig. 4.5. (bottom) An amphisbaenian (Amphisbaena alba) above and a dibamid (Dibamus novaeguineae) below. Both are relatively small (less than a meter long) burrowing lepidosaurs. A. alba lives in northeastern South America, and D. novaeguineae in New Guinea. (Not to scale.)

large marine lizards, now extinct, generally taken to be closely related to monitors. Lee's contention that they are the closest relatives of snakes was first proposed by Edward Drinker Cope in the late nineteenth century. Lee, using phylogenetic methods, concluded that monitors were also closely related to mosasaurs (but not as closely as snakes). His view that snakes were originally derived from marine squamatans was also previously proposed, this time by Nopsca. Thus, among living animals, monitors are the closest relatives of snakes, and the closest relatives of both groups taken together are the beaded lizards, the helodermatids (ignoring for the moment *Lanthanotus*). The recognition of a close relationship between varanids and snakes is of long standing, although Lee contends that with the application of phylogenetic methods the issue became doubtful. He argues that this was due to an inappropriate choice of groups with which to compare them,

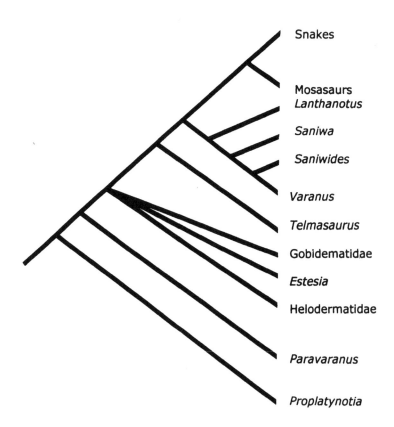

Snakes

Mosasaurs
*Lanthanotus*

*Saniwa*

*Saniwides*

*Varanus*

*Telmasaurus*

Gobidematidae

*Estesia*

Helodermatidae

*Paravaranus*

*Proplatynotia*

Fig. 4.6. The relationships of snakes, mosasaurs, and varanoid lizards, according to Lee (1997).

a conclusion that has not been unanimously accepted. Still, Lee has produced the most comprehensive analysis so far.

At this point, a brief digression into some phylogenetic concepts is appropriate. Phylogenetic systematics emphasizes relationships based on ancestry, as a child is related to her mother and grandmother, for example, rather than relationships based on resemblance, as my IBM clone is related to a real IBM PC. A group of animals that all share a common ancestor and that includes all the descendants of that ancestor is called a "natural" group. (Finding natural groups is important to phylogenetic systematists.) Examples of natural groups are birds or monitors—or, for that matter, people—but not reptiles, because the common ancestor of all reptiles is also the (remote) ancestor of mammals and birds. Here we refer to the traditional, or colloquial, notion of reptiles as including lizards, snakes, crocodilians, and turtles, not to any of the recent cladistic redefinitions of reptiles as, for example, consisting of lizards, snakes, crocodilians, and birds (lepidosauromorphs and archosauromorphs).

When we expand our horizons to include fossil as well as living lizards, some extinct groups turn out to be more closely related to varanids than to the other anguimorph lineages (xenosaurs and anguids). The group (clade) including the monitors, helodermatids, *Lanthanotus*, and the extinct groups (dolichosaurs, aigialosaurs, and mosasaurs)—and snakes—is the Platynota. The dolichosaurs, aigialosaurs, and mo-

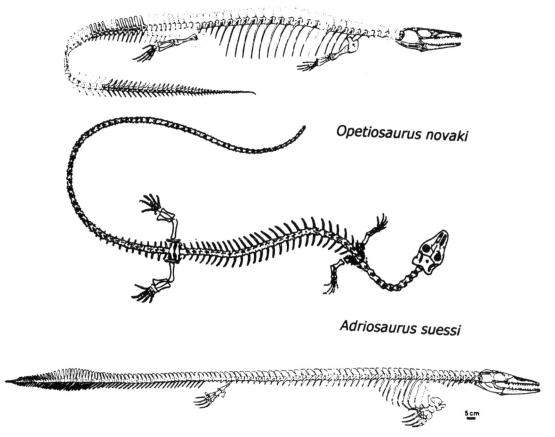

*Opetiosaurus novaki*

*Adriosaurus suessi*

5 cm

*Clidastes sternbergi*

Fig. 4.7. The reconstructed skeletons of an aigialosaur (Opetiosaurus), an dolichosaur (Adriosaurus), and a mosasaur (Clidastes). Not to scale: Opetiosaurus and Adriosaurus were approximately a meter or so long, but this skeleton of Clidastes was about three meters long, and other mosasaurs were considerably larger. (Modified after Romer 1956, and Caldwell et al., 1995.)

sasaurs (Fig. 4.7) were all marine (or at least aquatic) animals, and they all became extinct with the end of the Cretaceous.

Lee's analysis suggests that monitors probably arose sometime in the Early Cretaceous, between about 140 and 100 million years ago. It could not have been later because the dolichosaurs, moderately small marine lizards and the oldest close relatives of the mosasaurs (and hence with them relatives of the snakes), date from late in the Early Cretaceous, nearer 100 than 140 million years ago. Since dolichosaurs are also descendants of the ancestors of monitors, their existence indicates that this ancestral lineage must have split into two at the time: the dolichosaurs and the monitors. Otherwise, the dolichosaurs themselves would be the ancestors of monitors. Of course, this split could have happened before, and monitors could have arisen earlier, but so far no fossils clearly indicate that they did (more about that later).

### Ancient Relatives of the Monitors

A group of platynotan lizards is known from the Cretaceous, mostly from North America (Table 4). Such lizards also existed in Asia. We

Table 4
Fossil monitor-like lizards

| Periods (my = millions of years) | North America | Europe | Africa | Asia | Australia |
|---|---|---|---|---|---|
| Pleistocene (0–1.8 my) | | | | | |
| Pliocene (1.8–5 my) | | | | | |
| Miocene (5–23 my) | | | | | |
| Oligocene (23–34 my) | | *Necrosaurus cayluxi* (?) | | | |
| Eocene (34–55 my) | *Parasaniwa* sp. | *Eosaniwa koehni, Necrosaurus eucarinatus, Necrosaurus* sp. | | | |
| Paleocene (55–65 my) | *Palaeosaniwa* cf. *P. canadensis, Provaranosaurus acutus*, cf. *Provaranosaurus* sp. | *Necrosaurus* sp. | | | |
| Maastrichtian (65–71 my) | *Colpodontosaurus cracens, Palaeosaniwa canadensis, Parasaniwa wyomingensis* | | | | |
| Campanian (71–82 my) | *Palaeosaniwa canadensis* | | | *Estesia mongoliensis, Telmasaurus grangeri* | |
| Santonian (82–83 my) | | | | | |

know from continental drift that geography—the positions of the continents and oceans—was quite different in the past from that which we now learn in school. Early in the Cretaceous, around 130 million years ago, the landmasses were basically grouped into two supercontinents, Laurasia in the north (consisting of what later became North America and Eurasia) and Gondwanaland in the south (made up of South America, Africa, Antarctica, Australia, and India). By the Late Cretaceous, these supercontinents had begun to separate, but the terms "Laurasian" and "Gondwanan" are useful to indicate landmasses that

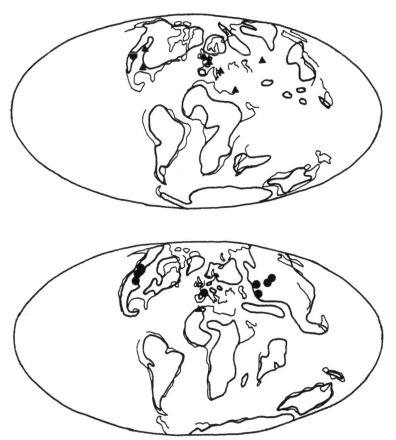

Fig. 4.8. Maps of Cretaceous platynotan fossil sites. Upper, Early Cretaceous; lower, Late Cretaceous. Circles indicate sites yielding land-dwelling platynotans; triangles, those yielding marine forms. Heavy lines mark Cretaceous shorelines (where known); light lines, modern coasts. Terrestrial platynotans were apparently restricted to the northern (Laurasian) lands.

still shared similar animals. In the European region—then an archipelago, a chain of islands like Indonesia is now—platynotans were represented by the dolichosaurs, and the oldest snakes are found along the shores of the continent just south of these islands, Africa. However, so far no traces of these animals have been found on any southern (Gondwanan) continent (beyond the northern shores of Africa, that is). Their presence on the Laurasian lands and absence from Gondwanan continents (Fig. 4.8) suggests that the platynotans evolved from their ancestors somewhere in Laurasia, and that monitors did so as well, only later migrating into both Australia and Africa from the north.

The oldest (known) platynotans come from the later part of the Early Cretaceous in North America. Until 1995 the marine platynotans of Europe were the oldest known. They clearly implied, since they were not ancestral to the later land-dwelling platynotans, that earlier platynotans had existed but hadn't been found. In 1995 a maxilla was found in the mid-Cretaceous Cedar Mountain Formation of Utah. This maxilla, together with some other cranial fragments and two vertebrae, was eventually described as representing *Primaderma nessovi* (Nydam 2000). This lizard flourished about 95 million years ago in company with such recently discovered dinosaurs as *Utahraptor* and *Cedarosaurus*. In view of the fragmentary remains, we do not know much

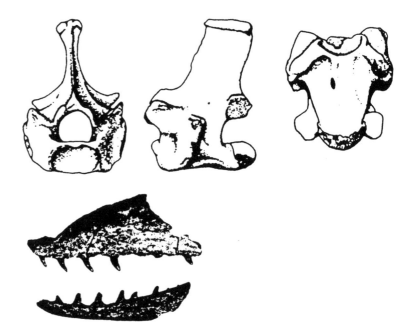

*Fig. 4.9. Fossils of* Necrosaurus cayluxi, *from the late Eocene of France. Maxilla and dentary below; vertebra in anterior, lateral, and ventral views above. (Drawn from de Fejervary 1935, and modified after Hoffstetter 1969.)*

about *Primaderma,* but it did have sharp, flattened, serrate, recurved teeth rather like those of later monitors. This suggests that already at this time, land-dwelling platynotans had become predators on relatively large prey (Nydam 2000) and thus had adopted the predatory habits that have stood monitors in good stead ever since.

However, *Primaderma* was not among the ancestors of modern varanids. It was mentioned earlier that the closest living relatives of monitors (including *Lanthanotus*), their sister group, are the helodermatids, the Gila "monster" and beaded lizard, among living lizards. The helodermatids, together with their extinct relatives, form a group known as the monstersaurs (Norell and Gao 1997). *Primaderma* is a monstersaur, the oldest known, and thus indicates that the origin of platynotans is older than the middle of the Cretaceous.

The oldest varanoids, the sister group of the monstersaurs, were the necrosaurs, a name indicating "lizards of death" (Fig. 4.9). Some necrosaurs (*Colpodontosaurus, Palaeosaniwa, Parasaniwa,* and *Provaranosaurus*) lived in North America during Late Cretaceous to Middle Eocene times (about 83–41 million years ago), and some (*Necrosaurus* and *Saniwa*) in Europe during the Late Eocene (about 41–38 million years ago). When alive, they doubtless looked much like existing monitors. They were also similar in the form of their teeth, with the basal "wrinkles" of infolded enamel that much later persisted in the teeth of *Megalania,* and in their relatively low, broad vertebral bodies (centra) in the vertebral column. Lee's analysis indicates that these animals were simply primitive forms no more closely related to each other than to the lineages that went on to become mosasaurs, snakes, helodermatids, and monitors—or, in the jargon, they did not form a natural

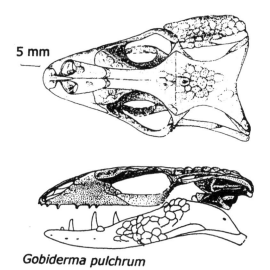

5 mm

*Gobiderma pulchrum*

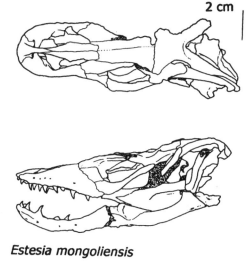

2 cm

*Estesia mongoliensis*

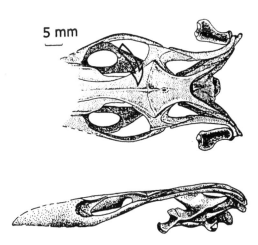

5 mm

*Telmasaurus grangeri*

*Fig. 4.10. (above) The skulls of two Mongolian Late Cretaceous monstersaurs,* Gobiderma *and* Estesia. *Both are shown in lateral (below) and dorsal (above) views. (Modified after Borsuk-Bialynicka 1984, and Norell, McKenna, and Novacek, 1992.)*

*Fig. 4.11. (left) The skull of* Telmasaurus, *from the Late Cretaceous of Mongolia, in dorsal (above) and lateral (below) views. The arrow indicates the midline crest (not visible in lateral view). (Modified after Borsuk-Bialynicka 1984.)*

group. (So there is no good reason to lump these forms together but exclude lineages like the monitors or the mosasaurs.) When alive, the Eocene ones may have been as much as a meter long, and *Eosaniwa* was possibly as long as two meters. The earlier forms (*Colpodontosaurus, Parasaniwa,* and *Provaranosaurus*) were smaller, probably less than a meter long. Judging from their teeth, they were all predatory lizards.

A visitor to the Late Cretaceous would have seen lizards that looked much like modern large monitors, since the body form—once achieved—changed little, while the anatomical details that distinguish modern varanids evolved. Forms such as *Estesia* and *Telmasaurus,* in Mongolia, would have looked familiar—at least to someone familiar with monitors. Both grew to a length of about one to two meters.

*Estesia* is known from an almost complete skull and jaws found in 1990 by a party from the American Museum in New York City (Fig. 4.10). Mark Norell, Malcolm McKenna, and Mike Novacek (1992), all from the American Museum, noticed that the teeth of *Estesia* had long grooves, similar to those of beaded lizards. They suggested that perhaps like beaded lizards, *Estesia* had poison glands. However, this suggestion has not been unanimously accepted. *Telmasaurus* (Fig. 4.11) is not as well known as *Estesia* since, unfortunately, it is represented by less complete specimens (Gilmore 1943; Borsuk-Bialynicka 1984). It had a low crest along the midline of the top of its skull (on the frontals), a feature to which we will return when discussing *Megalania*.

The oldest monitors known are middle Cretaceous in age, so it is generally thought that this group arose during the preceding Early Cretaceous or even earlier. Reports of earlier platynotans (or at least varanid-like lizards) include fossil forms dating well back into the Jurassic. *Paikasisaurus indicus* comes from the rocks of the Early Jurassic Kota Formation (approximately 210–190 million years old) of central India (Yadagiri 1986). It is thought by its describer to be related to *Parasaniwa*. Only two fragments of jaws (with teeth) and an ilium have been found, but the form of the teeth suggested varanid relationships. Although the fragments are very small, about 2 mm long, the crowns are flattened and show basal wrinkles. These characteristics are found in monitors but in few other lizards. However, a more recent study of *Paikasisaurus* showed that the basal "wrinkles" were not plicidentine (Evans, Prasad, and Manhas 2002). So given the very fragmentary condition of the fossils, there is no good evidence for relating it to the platynotans. In fact, it is not even clear that all of the fossil material derives from a single species.

Another possibility is the formidably named *Nuthetes destructor* from the latest Jurassic (or earliest Cretaceous) Purbeck beds of England, about 145 million years old (Owen 1854). Like *Paikasisaurus*, *Nuthetes* is known from a fragment of jaw with teeth, but this time a little larger, about 3.5 cm long (and with six teeth, rather than two). *Nuthetes* is now usually considered a small carnivorous dinosaur (theropod), although Owen thought it was related to the varanids, and he named it after the monitor lizard (from the Greek *nouthetitis*, meaning "monitor"). Originally the teeth were described as attached to the inside of the jaw (pleurodont), rather than set in sockets (as dinosaurian teeth are). However, later illustrations indicated that the teeth had a shallow trough extending up the crown from the base, just as in many theropod teeth (Owen 1884a). The indisputably pleurodont teeth of *Megalania* show no such feature, but instead an expanded, flattened base quite different from those of *Nuthetes*. This indicates that *Nuthetes* was, as most paleontologists believe, a small meat-eating dinosaur.

The Purbeck Limestone Formation that yielded the fossils of *Nuthetes* has also produced fossils of several lizards. Of these, *Parviraptor estesi* is rather what one would expect of an ancestral platynotan (Evans 1994). It was a land-dwelling predatory lizard with sharp, recurved (but apparently not flattened) teeth. Fossils of this lizard have been found in even older deposits in England, the Middle Jurassic Forest Marble, about 170 million years old. A second species, *Parviraptor gilmorei,* has been found in the Morrison Formation (Evans 1996), which has yielded fossils of such well-known dinosaurs as *Allosaurus, Diplodocus,* and *Stegosaurus.* Thus, *Parviraptor* was widespread on the northern continent (Laurasia) during mid-Mesozoic times.

The skull of *Parviraptor* was about 4 cm long, so it was not a particularly small lizard. The skull shows several character states—a palatine bone as wide as it is long, a broad interpterygoid vacuity, the sharp recurved teeth, and well-developed subolfactory processes on the frontals (among others)—that strongly suggest relationships with the platynotans. But it lacks plicidentine in the teeth, a posteriorly placed ascending process of the maxilla, oblique condyles on its vertebrae, and waisted centra that are characteristic of platynotans. However, these are all plesiomorphic states and thus do not eliminate *Parviraptor* from being ancestral to, or at least near the ancestry of, the platynotans. Thus, predatory monitor-like lizards seem to date back well into the times of the dinosaurs.

This is as far back as we can reasonably trace the ancestry of monitors. *Parviraptor* may well have been a proto-monitor, and *Paikasisaurus* may or may not have been related to platynotans—only further fossils will tell—but *Nuthetes* wasn't a lizard at all.

## History of the Monitors

The oldest "real" monitors—that is, varanids—come from the Late Cretaceous and include such animals as *Saniwa, Saniwides,* and maybe *Palaeosaniwa.* The most recent commentary on these creatures is that of Richard Estes (1983), in honor of whom *Estesia* was named. *Saniwa* (about one meter long) lived during the Eocene and Oligocene (about 55–23 million years ago) in North America and Europe. Like most of these creatures, it would have looked much like a living monitor but with a relatively larger head (Figs. 4.12, 4.13). *Saniwides* was an earlier form from the Late Cretaceous (about 80 million years ago) of Mongolia. Again, it would have looked similar to living varanids, but reportedly it had a very low and flattened snout, perhaps giving the head a kind of duck-billed appearance. *Palaeosaniwa* seems to be fairly poorly understood, but was large. A Late Cretaceous lizard from Alberta, Montana, and Wyoming, *Palaeosaniwa* was as large as "some of the largest species" of living monitors, according to Estes, so presumably about two meters long.

But probably the oldest varanid is represented by a single cervical vertebra (Fig. 4.14) from older Late Cretaceous beds (about 90 million years old) east of the Aral Sea in Kazakhstan (Kordikova et al. 2001). Although most likely a varanid of some kind, this bone may represent a more primitive (plesiomorphic) platynotan.

The fossil history of the living monitors (genus *Varanus*) leaves much to be desired, and when considering it, words like "incomplete," "fragmentary," or even "frustrating" spring to mind. Of twenty-seven specimens (excluding Australian material) recorded by Estes in his

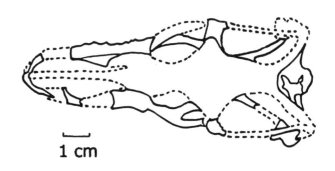

1 cm

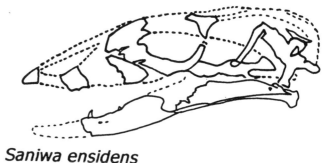

*Saniwa ensidens*

Fig. 4.12. (top) The skull of Saniwa, *a varanid of moderate size from Early Cenozoic North America and Europe. (Redrawn from Gilmore 1928.)*

Fig. 4.13. (center) Vertebrae of Saniwa ensidens, *from North America, and* S. orsmaelensis, *from Europe. The vertebrae are shown in ventral (left), lateral (center), and posterior (right) views. The posterior view of* S. ensidens *omits the transverse processes seen laterally in that of* S. orsmaelensis. *(Modified after Gilmore 1928, and Hoffstetter 1969.)*

Fig. 4.14. (bottom) Late Cretaceous vertebra from Kazakhstan, which may represent the oldest known varanid. The neural arch is missing, and the vertebra is seen in anterior (left), dorsal (middle), and ventral (right) views.

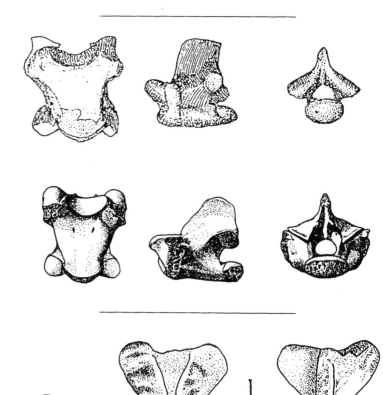

5 mm

**Varanoid from Kazakhstan**

1983 survey of fossil lizards, twenty-three included vertebrae, six had cranial elements, and only three had any limb material at all. In at least three of the specimens that included cranial elements, the dentary (the major bone of the lower jaw) was present. As we shall see, this pattern of preservation—mostly vertebrae, with perhaps more dentaries than expected but clearly fewer limb bones—is repeated in the Australian material.

The fossils of the "proto-monitors" like *Parasaniwa* and *Estesia* have been found in the Cretaceous rocks of North America and Asia (Table 4). The Early Cenozoic necrosaurs also lived in Europe. Thus, it is generally accepted that this group originated in the Northern Hemisphere. So far there are no fossils of these lizards from either Africa or South America.The fossil record of the genus *Varanus* begins just before the Miocene, and—outside of Australia—there is apparently a decrease in the occurrences of fossils with time, from the Miocene to the present. At first sight, this might suggest—as the scripts of some videos on oras claim—that monitors are a prehistoric form that has been slowly dying out with the passing ages. However, before concluding this, we need to see just how many occurrences of monitors there have been for equal durations of time.

The periods of time, called epochs, into which the Cenozoic has been divided reflect changes in the faunae and biological communities, not the passage of time as such; in other words, they are not all equally long. In fact, no two of them are of the same length. Thus, dividing the length (duration) of the epochs by the number of varanid specimens known (in practice, those listed by Estes 1983) should give a figure indicating how frequently monitor specimens are encountered in the fossil record. The Miocene lasted for 18.5 million years and has yielded eleven varanid specimens; the Pliocene, for 3.5 million years and has eight specimens; and the Pleistocene, for almost two (about 1.8) million years, with five specimens. Thus, for the Miocene there is approximately one monitor specimen per 1.7 million years; for the Pliocene, one per every half (actually 0.44) million years; and for the Pleistocene, one per every third (approximately) of a million years. Thus, monitors appear to have become more common recently than in the Miocene. Of course, the older the fossil record is, the more incomplete it is, so we should have some healthy skepticism about this result. But even so, it clearly does not show that monitors were more common during the Miocene, say 15 million years ago, than they are now.

The oldest fossils of *Varanus* itself date from about 25 million years ago (late in the Oligocene) in Kazakhstan, but these are very fragmentary. It interesting that the oldest likely varanid also comes from Kazakhstan. During the succeeding Miocene period, varanids were widespread, apparently living in most of the regions in which they still survive (Table 5). Fossils have been found in Europe, central Asia (Kazakhstan again), and east Africa. Bearing in mind that few if any fossil-bearing deposits of this age are known from India, Indochina, or Indonesia and that Europe has been more extensively searched for fossils than central Asia or east Africa, we can plausibly conclude that

## Table 5
### Fossil monitors

| Periods (my = millions of years) | North America | Europe | Africa | Asia | Australia |
|---|---|---|---|---|---|
| Pleistocene (0–1.8 my) | | | | Varanus hooijeri, Varanus bolkayi, Varanus komodoensis, Varanus sp. | Varanus varius, Varanus gouldii, Varanus cf. V. giganteus, Varanus emeritus, Varanus dirus, Megalania prisca |
| Pliocene (1.8–5 my) | | Varanus marathonensis, Varanus semjonovi, Varanus lungui | | Varanus sivalensis, Varanus darevskii | Varanus sp., Varanus dirus, Megalania sp. |
| Miocene (5–23 my) | | Iberovaranus catalaunicus, Varanus hoffmani, Varanus sp. | Varanus rusingensis, Varanus sp. | Varanus pronini | Varanus sp. |
| Oligocene (23–34 my) | Saniwa sp. | | | Varanus sp. | |
| Eocene (34–55 my) | Saniwa agilis, Saniwa brooksi, Saniwa crassus, Saniwa ensidens, Saniwa grandis, Saniwa paucidens, Saniwa sp. | Saniwa orsmaelensis | | | |
| Paleocene (55–65 my) | Saniwa aff. S. ensidens, cf. Saniwa sp. | | | | |
| Maastrichtian (65–71 my) | | | | | |
| Campanian (71–82 my) | | | | Saniwides mongoliensis | |
| Santonian (83–87.5 my) | | | | | |
| Coniacian (87.5–88.5 my) | | | | Varanid(?) | |
| Turonian (88.5–91 my) | | | | | |

*Note:* The specimen given as Coniacian may actually be Turonian in age, as the age of the beds in which it was found has not been precisely determined.

varanids were already widespread and thus that they have reasonably well held their own for the past 20 million years.

Also in the Miocene occurs the only monitor other than *Megalania* that does not belong to the genus *Varanus*. This is *Iberovaranus catalaunicus* from Spain and Portugal. Known only from its vertebrae, it differs from *Varanus* in having a more slender vertebral construction.

Most people who think about monitors at all—including a few who really should know better—think that there were only two giant monitors, oras and *Megalania*. But there was another, *Varanus sivalensis,* from the Pliocene rocks of the Siwalik Hills of northwestern India and adjacent Pakistan (Lydekker 1886). It is very poorly known and is represented only by two vertebrae and part of a humerus, but it is an important animal that we shall discuss later.

As for fossils of modern species, outside of Australia and Indonesia there is only a specimen somewhat tentatively identified as a Bengal monitor (*Varanus bengalensis*) from Pleistocene or Recent deposits in the Billa Surga Caves near Madras in India. Although it is nice to have the material, it tells us little: we could have guessed that modern species of monitors have been around for the past few tens of thousands of years.

## History of Monitors in Indonesia

Although there are no helpful Miocene deposits in Indonesia, there are more recent ones. And since this is the home of the ora, the varanid fossils from this part of the world are particularly interesting (Fig. 4.15). They have been found in Kalimantan, Java, Timor, Flores, and Irian Jaya (New Guinea). The fossils from Kalimantan, from the Malayan provinces of Sarawak and Sabah, represent unidentified (and hence presumably relatively small) monitors. They date from the Late Pleistocene or later, with the ages given as 900–22,000 and 2,000–40,000 years ago for two sites in Sarawak (Bellwood 1985; Datan 1993) and 10,000–15,000 years ago for the site in Sabah (Bellwood 1985). Recently, monitor remains, including "tooth-bearing elements," have been found in caves in the central Vogelkop (Bird's Head) Peninsula of Irian Jaya (Aplin, Pasveer, and Boles 1999). The deposits are believed to be Holocene and Late Pleistocene in age, and the fossils indicate "quite large monitors."

*Varanus hooijeri* is represented by parts of a skull from Recent cave deposits on Flores (Brongersma 1958). This animal had blunt, stout teeth, and Estes thought it was related to the Nile monitor (*Varanus niloticus*) or Gray's monitor (*Varanus olivaceus*).

The Pleistocene deposits at Trinil in Java, which have also yielded early human fossils, produced *Varanus bolkayi,* known from two long, slender vertebrae (de Fejervary 1935). The larger is 2.7 cm long, indicating a large, but not a very large, monitor. Another pair of vertebrae was found in 1966 in Pleistocene gravels in (western) Timor. Dirk Hooijer, at the Rijksmuseum in Leiden (the Netherlands), compared these vertebrae with those of a 2.5-meter-long skeleton of an ora and found them quite similar. Hooijer (1972) concluded that these, together

Fig. 4.15. Indonesia during the Pleistocene, with sites where varanid fossils have been found. The continuous lines indicate shorelines during the low sea levels; dotted lines indicate modern shores. Solid triangles indicate places where Pleistocene varanid fossils have been found, open triangles indicate where Holocene varanid material has been found, and the diamond indicates Komodo Island.

with fossil vertebrae from Trinil (those from *V. bolkayi* and others) and from Kedungbrubus, probably represented fossil oras. The deposits at both Trinil and Kedungbrubus are probably about 70,000–160,000 years old.

In 1981 Walter Auffenberg, a herpetologist at the University of Florida who conducted the most extensive field study of the ora, suggested that both *V. bolkayi* and the Timor material may actually be fossils of water monitors (*V. salvator*) rather than oras, although the reasons for this assessment have apparently never been published. However, fossil ora teeth have recently been reported from beds 850,000 and 900,000 years old on the island of Flores (Morwood et al. 1999; Sondaar et al. 1994). Oras have been present on Flores for at least half of the Pleistocene and may once have included Timor and Java in their range. It is clear there were monitors of (presumably) moderate size and also quite large ones—though it is not clear just how large—in Indonesia during much of the Pleistocene. Rumors of the discovery in Indonesia of fossils of very large monitors, of the size of *Megalania*, have circulated in the Australian paleoanthropological community. So far, however, these rumors have proved impossible to substantiate.

The number of species belonging to the genus *Varanus* is rather large, at least by tetrapod standards. Daniel Bennett, an expert on monitors at the University of Aberdeen (Scotland) and author of one of the few books (1998) on the subject, lists forty-five living species, and—omitting *Megalania*—there are at least nine extinct ones. The German herpetologist Robert Mertens was the first to attempt to seek order in this mass of species, and he proposed ten subgenera (which can be found in the appendix to Rodney Steel's 1997 book on monitors). Many herpetologists still use his classification today, but it did suffer from the fact that he worked during the Second World War and hence had access to only a limited supply of preserved specimens.

More recent work using biochemistry has been carried out by Australian herpetologists Susan Fuller, Peter Baverstock, and Dennis King (1998). They found that to a large extent the relationships of the different species accurately reflected their biogeography. In other words, the Australian monitors were generally all related, as were the Indonesian monitors and the Asiatic monitors. The African monitors, however, were not closely related to each other, and they were the most primitive. This supports the conclusion already mentioned that monitors evolved in the Northern Hemisphere and entered Africa early, possibly before arriving in Indonesia and Australia. So why is it not thought that monitors originated in Africa? Simply because the earliest platynotans, presumably including the ancestors of the monitors, were common in North America and Asia, and there is (yet) no indication of them in Africa. Fuller, Baverstock, and King also found that the ora is a recently evolved form, closely related to the lace monitor (*V. varius*) and the crocodile monitor (*V. salvadorii*), contrary to the idea proposed by Auffenberg that the ora was a primitive monitor.

These scientists found that the Australian monitors are all closely related. This implies that the pygmy monitors (subgenus *Odatria*) of northern and western Australia (and southern New Guinea) are not a natural group—some are more closely related to some of the large monitors of this region than to other species of *Odatria*. The three "Indonesian" forms are a second lineage and are the most closely related to the Australian lineage. The other Asian monitors (with the exception of the mangrove monitor, *Varanus indicus*) form a third.

The early view of herpetologists was that the ora evolved in Australia and later emigrated to Indonesia. The reasoning behind this view was never clearly stated (at least as far as I can determine), but it was apparently based on the ora's general similarity to the Australian monitors and the large number of those lizards found in Australia, including one species shared with Indonesia. Since biogeography is obviously important to monitors, and without going deeply into the biogeography of what is, after all, a complex region, we still need to try to understand that of the Malayan Archipelago (the southeastern Asian chain of islands). This is helpful in working out the history of the ora and its relationships—if any—to *Megalania*.

Large Asian animals inhabit the large islands of western Indonesia—Sumatra, Kalimantan, and Java—as well as smaller isles as far east as Bali. These were all part of the Asian mainland during low stands of the sea during the Pleistocene. Sulawesi (previously Celebes) was never connected to the mainland, although it is not far east of Kalimantan. This island is thought to have originated at the border of the Indo-

Australian plate well out to sea and to have moved northwestward to its present position. It has its own unique mammal fauna, including mostly rather small rodents and primates but also two species of pigs and two of buffalo. However, what captured most interest from biogeographers was that it had native marsupials, phalangers, as well. The islands further to the east, the Maluku (Moluccas) and Halmihera, for example, originated on the Indo-Australian plate and drifted in from the south. Before the arrival of people, these were probably inhabited only by marsupials (among mammals).

The Malayan Archipelago seems to be made up of three components. The western islands were joined to Asia from time to time, and so carried an Asiatic fauna. The eastern islands originated near Australia and New Guinea, and so carried an Australian/New Guinean fauna. And the intermediate islands, which may have arisen from the sea floor as the tectonic plates collided, were settled by migrants from both major landmasses.

In 1980 Auffenberg presented the results of geological research carried out by his team on Komodo. These indicated that volcanic eruptions in that region started at least 130 million years ago on both Komodo and Flores during the Early Cretaceous. Fossil wood showed that large trees grew there thereafter, and there was again volcanic activity in the Eocene, about 50 million years ago. Auffenberg concluded that although Komodo and Flores may have formed a single island early in their history, both have existed since Cretaceous times. Whether monitors have been there all that time is, of course, a different question, but the fossil teeth on Flores indicate that they were there by the Middle Pleistocene. If oras did originate in Australia, they could have arrived with the islands that would become the Maluku and the other eastern Indonesian isles. But an origin in Australia would imply that the ancestors of oras migrated from mainland Asia to Indonesia, then to Australia, and then back again to Indonesia—not impossible, surely, but in the absence of supporting evidence an unnecessarily complicated scenario.

Living oras eat fairly large mammals—deer, boar, goats, and the like. But these animals were all introduced to the Komodo-Flores region by people around 5,000 years ago, and oras have been there almost two hundred times that long (or longer). So what did oras eat before that time? Jared Diamond, a biologist at UCLA, suggested (1987) that they ate the native pygmy elephants, stegodonts (*Stegodon*), of these islands. Stegodonts were also found on Timor, where there were also large Pleistocene monitors, perhaps oras; and certainly large mammals were available as prey on Java. But pygmy elephants were not the only potential prey of Pleistocene oras. The giant tortoise *Geochelone atlas* had a range extending from the Siwalik Hills of Pakistan in the west to Flores in the east. Fossil shells of this animal from the Siwalik Hills reach 5 feet 6 inches (about 1.7 m) in length and 5 feet (about 1.5 m) wide—explaining why this animal was initially named *Colossochelys*. A fossil shell fragment from Flores, together with a pelvis and hind leg of one of these great tortoises was found associated with six ora teeth (Sondaar et al. 1994). These tortoises were certainly large enough to feed an ora, so oras may have enjoyed tortoise as well.

## History of Monitors in Australia

In Australia, as in Europe, monitors date back to Miocene times or perhaps a little before. These specimens, from deposits in the northern part of South Australia, are sufficient to identify that monitors were present—but no more. Varanid fossils are found in sediments of the Etadunna Formation exposed along the dry lakes of northern South Australia. These beds are at most about 25 million years old, from the Late Oligocene. As said before, this is effectively the beginning of the known fossil record for terrestrial animals in Australia.

Other monitor fossils have been found in Miocene beds in the Northern Territory (Estes 1984) and Queensland (Archer, Hand, and Godthelp 1991); in Miocene or Pliocene cave deposits in South Australia (Pledge 1992); in Pliocene beds in the Northern Territory (Tedford, Wells, and Barghoorn 1992) and Queensland; and in Pleistocene beds and cave deposits in Queensland, New South Wales, and South Australia. The early fossils are mostly vertebrae; in fact, so are most of the more recent fossils. But the more recent material also includes cranial bones, teeth, and a few limb elements.

It is clear that varanids as well as the other families of lizards have a long history in Australia. This does not imply that they belong to the groups of Gondwanan origin or preclude immigration from Asia. Although Australia approached Asia closely enough for rodents to enter through Indonesia only in the Pliocene, there is evidence that large islands, so-called "micro-continents," lay between Australia and Asia since early in the Cretaceous. These could have provided a route of access for lizards well back into the Paleogene, or perhaps even earlier.

With the Pleistocene material, better identifications can be made. The perentie (*Varanus giganteus*) is probably represented by fossils from Wellington Caves in New South Wales (de Fejervary 1918). Gould's goanna (*Varanus gouldii*) has been reported from deposits in Victoria Cave in South Australia, together with the lace monitor (*Varanus varius*) (Smith 1976). And further fossils, probably of Gould's goanna, have been uncovered in Henschke's Fossil Cave, also in South Australia (Pledge 1990). The perentie is no longer found anywhere near Wellington Caves. Since the modern range for the perentie is somewhat north and west of these localities, this suggests that they ranged further south and east during the warmer of the Pleistocene interglacials or during the warmer periods since the last ice age. Or perhaps their ranges have just contracted since people arrived in Australia. The remains from Wellington Caves come from two levels (which one yielded the perentie remains is not known)—one probably a little older than about 33,000 years ago, and the other probably between 14,000 and 3,000 years ago.

In 1995 my then-assistant, Mrs. Joanne Wilkinson, reported on two unusual tail vertebrae. The locality had not been recorded, but judging from the type of preservation, they were probably from the Darling Downs region. The caudals were unusual in the great height and vertical orientation of the neural spines. They were similar to those of the so-called water monitors such as Mertens's goanna (*Varanus mertensi*), and since such monitors had not previously been recorded as fossils, she published a short note on them (Wilkinson 1995). One problematic aspect, however, was that they also resembled the caudals of Gould's goanna. This was resolved a few years later by the biochemical work of Fuller and her colleagues (1998), who found that Mertens's goanna and Gould's goanna are closely related. Thus, these vertebrae further document the existence of this lineage in the Pleistocene or Pliocene.

## *Megalania* and Other Giant Monitors in Australia

A number of species of extinct large or giant fossil monitors have been described from Australia: *Varanus dirus, Varanus emeritus,* and *Varanus warburtonensis.* We might call these, respectively, the dire monitor, the venerable monitor ("emeritus" literally means "retired," but this hardly seems an appropriate name for a varanid), and the Warburton monitor. *Notiosaurus dentatus,* as mentioned in the previous chapter, is universally considered to be *Megalania.* The Warburton monitor is based on a moderately large vertebra from the Warburton River of South Australia—which, like the lakes of that region, is most often dry. As already mentioned, the describer of these remains, Etheridge (1894), thought they derived from an immature *Megalania;* but Zietz (1899) thought they represented a new monitor, hence the name. Subsequent workers have been unable to see any distinctive features of these fossils and so have unanimously agreed with Etheridge.

The dire monitor was named by Charles de Vis (1889) from an isolated tooth from the Darling Downs. He thought it was distinct from *Megalania* by virtue of its smaller size, and again, other paleontologists have been unable to find any reason to regard this as anything other than a tooth of *Megalania*—not the largest *Megalania,* to be sure, but still *Megalania.* But another specimen identified as the dire monitor (de Vis 1900) is more problematic. For one thing, it comes from the Pliocene deposits at Chinchilla rather than the Pleistocene beds further east on the Darling Downs. For another, it represents more than a single tooth: it is a maxilla with three teeth still in place. De Vis thought this element closely resembled that of the crocodile monitor, but it was about twice as large as that of the 7-foot (about 2.1 m) skeleton that he had available. However, it is considerably smaller and more lightly built than the well-preserved maxillae of *Megalania,* and therein lies the difficulty. Does this specimen indicate a distinct species of monitor (which lacks a name since *Varanus dirus* is *Megalania*), or is it from a juvenile *Megalania*? The problem is that there are no studies of the changes of the skeletal forms and proportions during the ontogeny—the growth to maturity—of any monitor. Most paleontologists have treated this as a specimen of *Megalania,* since it affords no clear reason not to, but with a definite feeling of unease.

The venerable monitor, *Varanus emeritus* (de Vis 1889), is also a problem, but in the opposite sense. It is unusual in that it is based on two pieces—both halves of limb bones—a humerus and a tibia. They are much smaller than those of *Megalania,* and while it is certainly possible that this lizard was a small and hence quite young *Megalania,* it is also possible that it was not. The proportions of the bones are unlike those of adult *Megalania,* but given the substantial difference in size, this may not be significant. On the other hand, I found these bones to be indistinguishable from those of the crocodile monitor (*V. salvadorii*), which is now found only in southern New Guinea. It is possible that this monitor could have moved from New Guinea to Australia across the land bridges that were present from time to time during the Pleistocene, or even moved the other way. So this material may represent that lizard or another like it.

Auffenberg (1981) argued that the ora outcompetes and hence drives out smaller monitors. If that was true of *Megalania*—and it clearly is not true of large living Australian monitors like the perentie (Pianka 1994)—then these pieces all may derive from small individuals of *Megalania.* But on the other hand, if they do not derive from *Megalania,* then they suggest that the biology of *Megalania* was different from that of the ora, perhaps more like that of the large Australian . goannas in this regard. The fossils of a lizard from the Mertens's-Gould's monitor lineage suggests that other monitors were indeed about. But since we don't know where these fossils came from, and whether *Megalania* was found there as well, we can't decide between these two possibilities with any confidence.

Mike Lee is the only person to have attempted to work out which living monitor is most closely related to *Megalania.* He has suggested (1996) that the perentie (*Varanus giganteus*) is the closest relative, a suggestion that makes sense on geographical grounds. Others had previously suggested that the ora is the closest relative but without giving any reasons for their conclusion (presumably it was based on size). Lee noted that one skull of the perentie bore an incipient midline crest on the skull roof similar to that of *Megalania* (discussed in the next chapter). I have also seen this kind of incipient crest on other perentie skulls. Unfortunately, the function, if any, of this feature in perenties is unknown—it seems not to be apparent in live animals. Lee suggested that the lineage of the perentie and *Megalania* had recently evolved this feature and thus that it indicated a close relationship (was a "shared, derived feature" in the jargon). However, he was unaware that such incipient crests are also found in several modern Australian species (Molnar 1990) and that they date back a long way among monitors and their kin. As we have seen, it was also found in the Cretaceous *Telmasaurus.* So although *Megalania* and the perentie may be related, the same evidence could be used to support a relationship to the lace monitor (*Varanus varius*), the plains monitor (*Varanus spenceri*), or the mangrove monitor (*Varanus indicus*).

Max Hecht pointed out that there was a second but undefined species of *Megalania* in addition to the Pleistocene *Megalania prisca.* This animal is known only from the Pliocene deposits at Chinchilla, which lie some 75 km northwest of the fossil-bearing soils of the eastern Darling Downs, on the western Darling Downs. The beds at Chinchilla have long been thought to be older than those of the eastern

Downs. Recent work comparing their fossil fauna with that from the early Pliocene deposits of the Burdekin River in northeastern Queensland—which have been radiometrically dated as about 4.5 million years old—indicates that they are indeed of approximately the same age. Hecht noted that this *Megalania* was represented by several vertebrae, generally smaller than those from the eastern Downs and other Pleistocene deposits (but not smaller than all of them). However, local naturalist Malcolm Wilson collected the single limb bone from Chinchilla, a fibula. This bone is 21 cm long, very nearly as big as the only other known fibula (23 cm) from the Pleistocene deposits—and that from Chinchilla is broken at its distal end. Given that more fossils are known of the Pleistocene *Megalania* and therefore large individuals are more likely to be represented, this size suggests that the Pliocene *Megalania* was probably already as big as *Megalania prisca*.

The date of extinction of *Megalania* is not clear. Few Pleistocene deposits in Australia have been dated, and, as we have already observed, *Megalania* is not a common fossil. Although found at Wellington Caves, it has not been found in the dated levels, most of which are less than 33,000 years old. The archaeological excavations at Cuddie Springs in northern New South Wales also encountered fossils of *Megalania*. According to one report (Dodson et al. 1993) they were found in "unit 7"; according to the other (Furby et al. 1993), in "unit 8." Being archaeological excavations, the units were dated, but the oldest datable unit was "unit 6" at about 30,000 years ago. The two reports may not agree, but both attribute the *Megalania* remains to beds older than 30,000 years, and at least this is consistent with their occurrence at Wellington Caves in units older than 30,000 years. It would be interesting to know just when, prior to 30,000 years ago, *Megalania* became extinct—but we don't. The impressive work of Roberts and his colleagues (2001) concentrated on large marsupials and the giant flightless bird *Genyornis* (which was also dated by Miller et al. 1999). Roberts and his colleagues found evidence that the beds at Cuddie Springs had been disturbed, so the previously published dates may not be reliable. But the *Megalania* fossils derived from layers too old to date in any case. Rather younger dates for *Megalania* (19,000–26,000 years) have been published (Pianka 1994) but without the supporting data that would enable one to find out if the dates had been obtained with due regard for the current protocols. These ensure a greater degree of accuracy than for dates obtained prior to about 1985 (Meltzer and Mead 1985; Baynes 1999). Although these dates are probably incorrect, and although *Megalania* probably died out around 46,000 years ago, the possibility that it survived the megafaunal extinctions for a few millennia should not be ruled out. It was a formidable creature that could discourage people from interfering with it, and (as we shall see in chapter 6) it may have been able to get by eating the larger of the surviving kangaroos. (This in no way, however, implies that I think it still survives—alas.)

Although most bones of *Megalania* have been mineralized to some extent, or at least stained by minerals dissolved in the groundwater, one specimen in the Queensland Museum was not. These two pelvic bones were so dry and chalky that I had them impregnated in resin to reduce the risk of their breaking up. They gave the impression of being only a few centuries old at most. Perhaps they were—but impressions can be misleading. I have seen and excavated bones in Montana that similarly appeared to be only a few hundred years old but that actually came from dinosaurs that died about 150 million years ago. The appearance of the bone may tell a lot about what minerals it was (or was not) exposed to in the groundwater, but it says disappointingly little about how old the specimen is.

## The Fossils of *Megalania*

Although *Megalania* is not an exceptionally difficult animal to understand—at least not once its remains have been properly sorted from those of *Ninjemys*—little has been known about the creature. As so often is the case with fossils, the problem is their absence. As intimated previously, most of the specimens of *Megalania* are vertebrae. In the Queensland Museum—which holds the most comprehensive collection of *Megalania* fossils—there are approximately one hundred dorsals, about half as many caudals, and five sacrals or parts of sacrals. These are approximate numbers because, although the great majority of the vertebrae are in quite good condition, there are some fragments, and it is not possible to be sure how many vertebrae these fragments really represent. However, we can be certain that there are slightly more than 100 dorsals, but probably not more than 110, and more than 50 caudals, but probably not more than 60. There are also eleven pieces from skulls or jaws, one piece of shoulder girdle, parts of three pelvic girdles, four bones (or parts thereof) from the forelimb, and also four bones from the hindlimb. Finally, there is one incomplete limb bone that I was unable to identify. Most limb bones are known from only a single specimen of each, the exceptions being the ulna and fibula, and so far as I can tell, there are no radii at all. One may be forgiven for thinking that the fossils of such a large animal should be easier to find.

The vertebrae tend to be well preserved and complete, whereas the limb and girdle elements tend to be broken and hence incomplete. About half the cranial bones are complete and half incomplete, and the incomplete ones are those of the upper and lower jaws. These are more lightly constructed than those from the back of the skull, which are massive, thick elements so far as we know. The vertebrae, also, are generally robust, compact bones, lacking projections that could break off—except the neural spines, of course, which often did break. On the whole, the preservation of these bones holds no surprises: those that were massive and without projections tend to be well-preserved, and those that were less strongly constructed or with projections tend to be broken.

There are—roughly—fifteen times as many vertebrae as bones from the skull or from the limbs and limb girdles. Of the specimens of non-Australian fossil monitors that Estes (1983) listed, there were about four times as many with vertebrae as with cranial bones, and almost eight times as many with vertebrae as with bones from the limbs. This is a bit like comparing apples with oranges since we are comparing, in the one case, individual bones, and in the other, specimens that presumably represent individual animals. Nonetheless, it is clear that the trend is the same for both. It seems that for monitors in general, the vertebrae preserve better than the bones of the skulls or limbs—the latter perhaps faring the worst. This is probably due to the differences in the construction of the bones and also because any individual animal has more vertebrae than other bones. What is odd—and I have no explanation for this—is that there is only a single neck (cervical) vertebra of *Megalania* in the Queensland Museum, presumably the one that Hecht reported finding when he examined all the collections holding specimens of *Megalania* almost thirty years ago. He thought the absence of cervicals might well be due to their weaker construction.

The dorsal vertebrae of living monitors—I checked the lace monitor (*Varanus varius*)—are all about equal in breadth (within 10 percent), so we can estimate how many individuals are represented by the dorsals in the Queensland Museum by measuring their breadth. Combining this with data on where they were found, we can arrive at an estimate of the minimum number of individual animals represented. This comes out to be twelve. Hecht reports material from a further seven sites, mostly outside Queensland. Including a more recent discovery near Townsville in north-central Queensland, and assuming there was only a single individual represented at each site, we get a minimum number of twenty *Megalania* represented in museum collections.

## Is *Megalania* a Valid Genus?

Herpetologists who have studied monitors have resolutely maintained only two genera: *Varanus,* and the early extinct *Iberovaranus.* The argument supporting this paucity of genera is that varanids are very conservative in their anatomy. So monitors have been divided into a set of subgenera rather than genera. Conservatism of anatomy is not a convincing argument in phylogenetic systematics. In that view, some objective criterion for delineating a genus, such as geological age, is preferred. If this approach were applied to monitors, the subgenera would be elevated to become genera.

The problem for *Megalania* is that there is no recent comprehensive study either of its anatomy or of its phylogenetic relationships. In fact, it is not clear that there is sufficient fossil material to carry out such a study. *Megalania* certainly has some unusual features, but these are of unknown significance, and many, although certainly not all, may be simply the result of its large size. Many herpetologists have suggested that *Megalania* should be renamed '*Varanus priscus,*' but they then refrain from actually doing so formally. Although I have used the name *Megalania* and will continue to do so, this name will not stand unless a phylogenetic analysis—when one is finally done—proves *Megalania* to be more distinct than it appears.

Traditionally, related species were placed in separate genera if the differences between them were immediately obvious. If they differed by having twelve scales, say, beneath the eye rather than sixteen, they were considered different species. But if, for example, one species had a crest of spines down the back and the other lacked this, they were considered to belong to different genera. Lizards, like other organisms, evolved in certain habitats, and their evolution was determined by which survived to successfully reproduce and how they were able to do this. For any single habitat, there are many ways in which this can be done. Thus, there are no inherent evolutionary factors that result in one set of features automatically, so to speak, distinguishing between genera, and another distinguishing between species (belonging to the same genus). So in the traditional approach, there was no way to objectively distinguish genera. For species, the situation was different—at least in theory. Species are groups of interbreeding individuals. In other words, any male of a species should be able to successfully mate with any female and produce offspring. Although this may be a satisfactory way of determining species for lab or zoo animals—or even large and easily observed animals in the wild—it proved very much less than satisfactory for small, shy, rare, or simply "hard-to-get-to" species. In practice, most animals are still attributed to species on the basis of their physical features rather than on whether or not they can successfully mutually reproduce. And of course this criterion of reproduction is impossible to apply to extinct animals.

To the phylogenetically inclined, this was an unsatisfactory situation—species could (in theory!) be defined objectively, but not genera or any other higher categories. The proposed answer was to define genera as species or groups of species that had been separate from other comparable evolving lineages for longer than some given period of time. A good phylogenetic analysis should be able to discern how long species have been separate and so automatically distinguish genera. So far, this criterion has not been generally accepted by taxonomists, phylogenetic or not. It could be applied to *Megalania* and its status determined. In this case, it would seem likely that the current subgenera of *Varanus* would become genera.

So does *Megalania* have sufficiently distinct features, or has it been separate from the other species for long enough, to warrant remaining a valid genus? So far, we don't know. *Megalania* has probably existed for at least 4.5 million years (since the Early Pliocene): if this is long enough to mark out a genus, then it will. If the current subgenera of monitors are retained as subgenera, then—unless it proves to be more unusual than it now appears—*Megalania prisca* will become *Varanus priscus*. However, even if the subgenera merit generic status, the name *Megalania* will probably still become obsolete.

It seems likely that *Megalania* will prove to be one of the Australian radiation of *Varanus*, the subgeneric name of which is also *Varanus* (since, according to the guidelines of taxonomy, one subgenus has to have the same name as the genus). If the subgenera are made genera, the subgeneric names become the generic names, so it would still be *Varanus priscus*. We can, of course, still use "megalania" as the vernacular name for *Varanus priscus*.

# 5. The Paleobiology of *Megalania*

## *Megalania* Alive

One goal of paleontology is to visualize, and to understand as living creatures, animals now represented only by fossilized fragments. The importance of creatures in the evolution of life lies in how they lived and died, and only by understanding that do we gain insight into their place in the history of life.

To do this, we must consider *Megalania* as living creatures. But we cannot just imagine them; we need some way of working out what is a plausible visualization and what is not. The demands of life provide some constraints for our imagination. All living creatures have to obtain energy—food, if you are an animal—to remain alive. They must be swift enough to catch their prey (if predators) and to escape their predators (if prey), at least on average. They cannot flout the laws of physics in their construction: their bones and teeth must be strong enough to resist the usual stresses imposed on them—"usual," for bones and teeth do break on occasion. They must be able to find mates and raise young. *Megalania,* like other lizards, used sex for reproduction and the adults probably did not die before the young were hatched, so social interactions were possible, both those between the sexes and those among parents and offspring.

In addition to looking at the demands on living creatures in general, we can look at oras for some guidance. *Megalania* were similar to the oras in form, and their habitats were also probably similar in terms of climate and plant communities (although the individual plants were probably different). Putting all this together, we can try to work out how *Megalania* lived. There will be gaps and uncertainties in our attempt, but if nothing else, it will introduce us to the problems of life faced by *Megalania*.

Table 6
Estimates of the size of *Megalania*

| Author | Date | Length (metres) | Length (feet) | Weight (kg) | Weight (lbs) |
|---|---|---|---|---|---|
| de Vis | 1885 | 4.6–5.5 | 15–18 | - | - |
| Lydekker | 1888 | 9.1 | 30 | - | - |
| Owen (in Zietz) | 1899 | 6.1 | 20 | - | - |
| von Huene | 1956 | 5 | 16.4 | - | - |
| Hecht | 1975 | 7 | 23 | 600–620 | 1,323–1,367 |
| Museum of Victoria skeletal replica | c. 1975 | 5.7 | 18.7 | - | - |
| Rich and Hall | 1979 | 5.5 | 18 | - | - |
| Auffenberg | 1981 | 4.5 | 14.8 | 2,200 | 4,850 |
| Bakker | 1986 | 4.6 | 15 | 453.6/1,814.4 | 1,000/4,000 |
| anonymous | 1986 | 7.5 | 24.6 | 620 | 1,367 |
| Paul | 1988 | - | - | 500 | 1,102 |
| Flannery | 1994 | 7 | 23 | 500+ | 1,102+ |
| Paul | 1995 | c. 6 | c. 19.7 | - | - |

For Auffenberg, the length is not a maximal estimate, but that for which he estimated the weight.
Except for Hecht, who clearly states which is maximal and which are other estimates,
I presume the other authorities give a maximum estimate.

## Estimating the Length

The single most striking aspect of *Megalania* is its size. It is clearly the largest land-dwelling lizard known and is impressively bigger than the ora. Unfortunately, it is not clear just how big *Megalania* really was. As already mentioned, there are no complete skeletons, not even a half complete one. This has not stopped anyone from estimating the size of *Megalania*, and in fact, there are quite a few estimates as shown in Table 6. But how seriously should we take these when competent scientists, working with similar material, estimate the total length as being anywhere from 4.6 to about 9 meters? Physicists have been heard to say disparagingly that measurements made by astronomers are accurate to

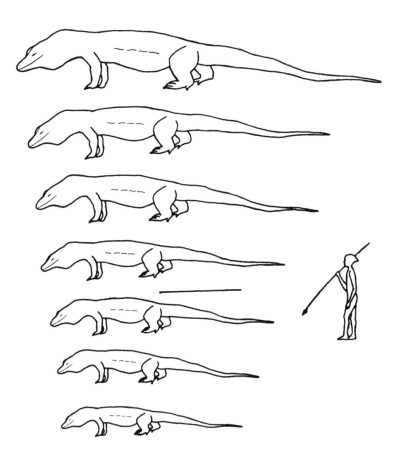

*Fig. 5.1. Various estimated lengths of* Megalania, *given as total length. From the top: 9.1 meters by Lydekker (1888); 7.5 meters by anonymous (1986); 7 meters by Hecht (1975) and Flannery (1994); 6 meters by Owen (in Zietz 1899: actually 6.1 meters) and Paul (1995); 5.5 by meters Rich and Hall (1979); 5 meters by de Vis (1885: given 4.6–5.5 meters) and von Huene (1956); and last, 4.5 meters by Auffenberg (1981) and Bakker (1986: actually 4.6 meters). The estimates of Hecht (1975) and Flannery (1994) represent the largest size for which there is adequate evidence. See Table 6. The restoration used here was inspired by a photo of* Varanus giganteus *in Bennett (1998).*

give or take 100 percent (which is true, incidentally, for objects beyond our galactic neighborhood). So is this range of estimates. None of the paleontologists have published how they arrived at their estimates. Presumably, they compared the dimensions of the bones of *Megalania* with those of the same bones of living varanids—but modern monitors do vary in their proportions, so the choice of which monitor to use affects the estimate. Oras are relatively short and stocky, while the crocodile monitor (*Varanus salvadorii*) is long and slender with a proportionately long tail. Max Hecht has argued that because oras have relatively short tails, so too might have *Megalania.* Sir Richard Lydekker, at least, could not have taken this into account in his estimate because oras had yet to be discovered at the time. So he must have used a more slender form with a relatively longer tail as a model. The various estimates in the literature are summarized in Table 6 and Fig. 5.1.

Hecht's estimate of 7 meters (total length) is based on a single claw (ungual phalanx) that appears to have been misidentified. This claw (UCMVP 56420) differs in form from those of modern varanids and may actually be from a bird. Hecht himself was doubtful that it really belonged to *Megalania,* a point that escaped most of his readers. Hecht believed the material collected by Lau (now in the Australian Museum) and Frost (now in the Queensland Museum) derived from animals 3.3–

Table 7

Hecht's (1975) estimates of length of individuals of *Megalania*
(assuming total length = 1.5 x body length)

| Bone | Specimen | Body length (SVL) (meters) | Body length (SVL) (feet) | Total length (meters) | Total length (feet) |
|---|---|---|---|---|---|
| Humerus | QM F865 | 1.5–1.6 | 4.9–5.2 | 2.25–2.4 | 7.4–7.9 |
| Ribs, ulna, jaw | Frost | 2.2–2.4 | 7.2–7.9 | 3.3–3.6 | 10.8–11.8 |
| Ribs, ulna | Lau | 2.2–2.4 | 7.2–7.9 | 3.3–3.6 | 10.8–11.8 |
| Dentary | QM F6562 | 3.1 | 10.2 | 4.65 | c. 15.3 |
| Vertebra | QM F2947 | 3.8 | 12.5 | 5.7 | 18.7 |
| Ungual | UCMP 56420 | 4.5 | 14.8 | 6.75 | c. 22.2 |

Presumably the estimates for the Frost and Lau specimens were based on the ulnae.
I believe that the specimen represented by the ungual is not *Megalania* and hence
that this estimate is fallacious.

3.6 meters long. He estimated that a dentary in the Queensland Museum (QM F6562) came from an individual 4.65 meters long and that the largest known bone, a vertebra (also in the Queensland Museum), was from an animal 5.7 meters long. This vertebra represents the largest known *Megalania* and would have come from a skeleton the size of that reconstructed by Tom Rich. (Hecht's estimates are summarized in Table 7.)

However, as luck would have it, the largest vertebra posed a minor problem. Hecht cited the specimen as QM F2947, but this specimen is a diprotodont ulna. The largest vertebra I could find was QM F2942, which is presumably the bone Hecht examined—after all, the terminal 7 could easily be mistaken for a 2. He gave the length of the body of the vertebra (centrum) as 66.5 mm, and the length of the centrum of QM F2942 is 66.5 mm. From personal experience I know that it is easy enough to make an error in recording specimen numbers, but it is also quite possible that the error was introduced during the process of publication.

My own estimates are based on comparing the size of the bones of *Megalania* with those of a living species of monitor (Fig. 5.2). A 1.24-meter-long skeleton (QM J16156) of the lace monitor (*Varanus varius*) was chosen, not because this animal is thought to be especially like *Megalania,* but for the very practical reason that the skeleton was the most complete and in the best condition. I measured the lengths of

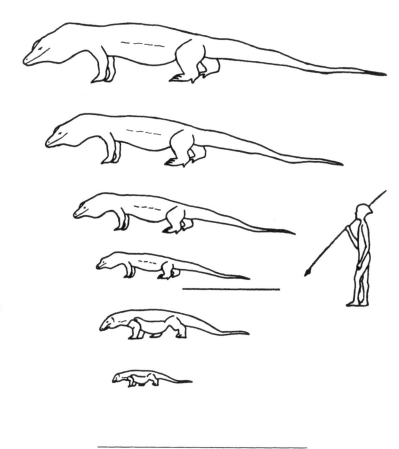

Fig. 5.2. (top) The estimates of the length of Megalania derived here, given as SVL. From the top: 3.8 meters for QM F2947; 3.1 meters for QM F6562; 2.3 meters for the Lau/Frost specimen(s); and 1.6 meters for QM F865. Below these are shown two oras, the upper at 3.09 meters (total length) the largest recorded, and the lower the average size given by Auffenberg (1981) of 1.7 meters. See Table 8.

Fig. 5.3. (bottom) The vertebral dimension used in estimated the size of Megalania. Reasons for using this particular measurement are given in the text.

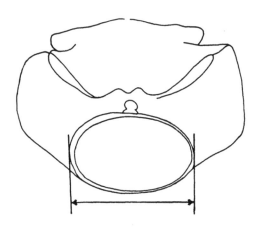

Table 8
Relative proportion of the tail in living species of *Varanus*,
based on data in Bennett (1998)

| Species | Total length | Snout-Vent Length | % of total length made up by tail |
|---|---|---|---|
| V. salvadorii | 251 | n.g. | 70 |
| V. glauerti | 80 | 30 | 69 a |
| V. glebopalma | 100(+) | n.g. | 67.5 a |
| V. mitchelli | 70 | n.g. | 65.5 a |
| V. pilbarensis | 50 | n.g. | 65.5 a |
| V. baritji | 72 | 25 | 65 |
| V. gouldii | 137 | 50 | 64 |
| V. beccarii | 94.5 | 34 | 63 |
| V. jobiensis | 120 | 45 | 63 |
| V. eremius | 46 | 17 | 63 |
| V. storri | 40 | n.g. | 63 a |
| V. bengalensis | 160 | n.g. | 61 |
| V. niloticus | 188 | 74 | 61 |
| V. timorensis | 60 | n.g. | 61 |
| V. varius | 198 | n.g. | 61 |
| V. indicus | 140 | n.g. | 60 |
| V. olivaceus | 176 | 73 | 60 |
| V. salvator | 200 | 80 | 60 |
| V. acanthurus | 60 | 25 | 58 |
| V. mertensi | 126 | n.g. | 58 |
| V. griseus | 120 | n.g. | 57.5 a |
| V. caudolineatus | 32 | n.g. | 55 |
| V. flavescens | c. 100 | n.g. | 55 |
| V. giganteus | 202 | n.g. | 55 |
| V. gilleni | 40 | 18 | 55 |
| V. semiremex | 60 | 27 | 55 |
| V. yemenensis | 110 | 50 | 52–55 |
| V. albigularis | 126 | 61 | 52 |
| V. spenceri | 125 | n.g. | 51–52 |
| V. telenesetes | 42.5 | 21.7 | 49 |
| V. brevicauda | 23 | 12 | 48 |

a: average value of proportion; n.g.: measurement not given.

various limb bones (humerus, ulna, femur, and fibula) as well as the maximum width of the dorsal vertebral body across the anterior articular face (Fig. 5.3) and compared these to the total length (as preserved, a few of the terminal tail vertebrae were missing). Some vertebrae of *Megalania*—where the anterior face was not well preserved—were measured across the posterior face. The breadth of the posterior face differs by less than 10 percent from that of the anterior. Again, the vertebral diameter was chosen for three practical, rather than theoretical, reasons: this dimension is easy to measure; the vertebral bodies are commonly broken at one or the other end, so the length of the body cannot be accurately measured but the breadth can; and vertebrae (particularly dorsals) are the most commonly preserved elements of *Megalania,* so this dimension gives the greatest range of data. Because the dorsal vertebrae vary slightly in size along the back, and because the exact position of those of *Megalania* is not known, the average vertebral diameter was used.

Other skeletons of living species of monitors in the Queensland Museum were also measured to ensure that the width of the dorsals did, in fact, reliably indicate the total length of the skeleton. And this does seem to be the case. So the width of the dorsals indicates the total length of the skeleton to (probably) within about 5 percent.

A problem with this method is that living monitors do vary somewhat in proportions: some have relatively longer tails, and others relatively shorter (Table 8). We have no idea whether *Megalania* might have had a relatively long tail or a relatively short one, but Hecht suggested that it was relatively short, half or a third of the body length (Fig. 5.4). Herpetologists get around the variation in tail proportions by using, not the total length, but rather the snout–vent length (abbreviated SVL), that is, the distance from the tip of the snout to the cloaca (Fig. 5.5). This I take to be the length Hecht referred to as the "body length." Skeletons show the total length, but the snout–vent length must be estimated. This can plausibly be done because the position of the cloaca (vent) in relation to the bones of the pelvis is known; it is placed just behind the ischium. For skeleton QM J16156, the snout-vent length was estimated at 0.54 meters, or 51.67 times as long as the average width of a dorsal vertebra. For this specimen, the tail is about 1.3 times the length of the body, a point that will be of interest later.

To a nonscientist, this value, 51.67, looks satisfyingly precise, as expected of a scientific measurement. But a physicist would notice that this is only about 3 percent more than 50 and wonder if measurements of enough specimens of monitors would show that the average value is 50. Without measuring fifty or a hundred skeletons—that in any case were not available—there are several issues to consider. First, the snout–vent length is itself estimated and is accurate to (at least) one place but (almost certainly) not to four. Second, the figure used for the vertebral diameter was an average, and no dorsals actually had that diameter. Third, individual animals vary in their anatomy, particularly as the result of variation in the availability of food as they are growing and developing. Thus, the form of the skeleton and of the individual bones is a compromise between what is specified by the genes (which is

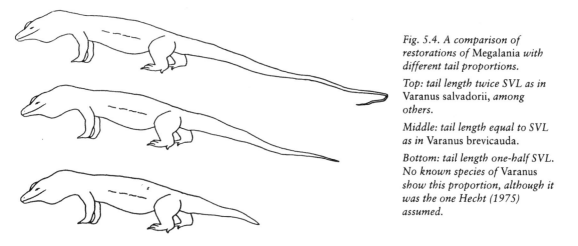

Fig. 5.4. A comparison of restorations of Megalania *with different tail proportions.*

Top: *tail length twice SVL as in* Varanus salvadorii, *among others.*

Middle: *tail length equal to SVL as in* Varanus brevicauda.

Bottom: *tail length one-half SVL. No known species of* Varanus *show this proportion, although it was the one Hecht (1975) assumed.*

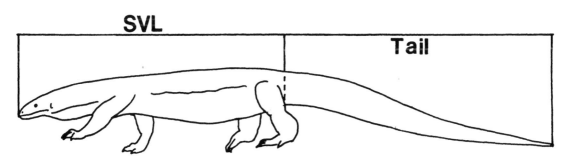

Fig. 5.5. Snout–vent length (SVL).

almost certainly not their form as such, much less the ratio of the average diameter of the dorsals to the snout–vent length) and the input of nutrients. Furthermore, it is well known, particularly for humans (e.g., Kennedy 1989), that bones respond to the stresses to which they are subject during life, which in turn depend upon the level and kind of activities engaged in by the individual, and that this also affects their form and development. All of these considerations suggest that while 51.67 is an appropriate factor for the lace monitor skeleton actually measured, other individuals almost certainly would have slightly different values. Thus 50 probably gives a reasonable estimate of the snout–vent length without conferring a false sense of accuracy.

My estimates differed by 5 to 10 percent from those of Hecht (Table 9), but given the sources of uncertainty, they can be considered to agree. The smallest *Megalania* dorsals in the Queensland Museum are about 12 mm in width (the smallest is 11.8), and the largest is 68.8 mm wide. These correspond to snout–vent lengths of 0.6 and 3.44 meters. Many *Megalania* dorsals (56%) are between 28 and 58 mm in width, which would correspond to lizards with snout–vent lengths of 1.4 and 2.9 meters.

Table 9
Comparison of estimates of Hecht (1975) and those made for this study.

| Specimen | Hecht's estimate | | Estimate carried out here | | % difference |
|---|---|---|---|---|---|
| | Specimen number | Snout-Vent Length | Specimen number | Snout-Vent Length | |
| Occiput | BMNH 39965 | 2.9 | - | - | - |
| Dentary | QM F6562 | 3.1 | - | - | - |
| Dorsal | "QM F2947" | 3.8 | QM F2942 | c. 3.5 | c. 10 |
| Dorsals | not given | 2.2–2.4 | - | - | - |
| Sacral | UCMP 56423 | "size of very large ora" | - | - | - |
| Humerus | QM F865 | 1.5–1.6 | - | - | - |
| Ulna | AM F2207 | 2.0–2.3 | - | - | - |
| Ulna | QM F867 | 2.0–2.3 | QM F867 | c. 2.5 | c. 9 |
| Radiale | AM F2208 | 2.3–2.4 | - | - | - |
| Ilium & pubis | QM F877 | 1.4–1.6 | - | - | - |
| Femur | AM F2206 | 2.2 | QM F4452 | c. 2.1 | c. 5 |
| "smaller terminal phalanges" | not given | 2.5 | - | - | - |
| "larger phalanx" | not given | 3.0–3.5 | - | - | - |
| "second largest phalanx" | not given | 2.6–2.7 | - | - | - |
| Ungual | UCMP 56470* | "nearly" 4–5 | - | - | - |

*Note that ungual UCMP 56470 is here regarded as probably not from *Megalania*: the details are given in the text.

## Table 10
### Estimates of length of individuals of *Megalania* assuming three different ratios of total length (TL) to body length (SVL)

| Bone | Specimen | Body length (SVL) (meters) | Total length (meters) for TL =1.5 SVL | Total length (meters) for TL = 2 SVL | Total length (feet) for TL = 2.5 SVL |
|---|---|---|---|---|---|
| Humerus | QM F865 | 1.5–1.6 | 2.25–2.4 | 3.0–3.2 | 3.75–4 |
| Ribs, ulna, jaw | Frost | 2.2–2.4 | 3.3–3.6 | 4.4–4.8 | 5.5–6.0 |
| Ribs, ulna | Lau | 2.2–2.4 | 3.3–3.6 | 4.4–4.8 | 5.5–6.0 |
| Dentary | QM F6562 | 3.1 | 4.65 | 6.2 | 7.75 |
| Vertebra | QM F2942 | 3.8 | 5.7 | 7.8 | 9.5 |

To see how much the proportions of the tail affect the result, consider the largest individual (QM F2942). (See also Fig. 5.4 and Table 10.) Using Hecht's notion of a tail one half of the snout–vent length, this individual would have been about 5.2 meters long. If the tail were equal to the snout–vent length, it would have been about 6.9 meters long; and if, like the lace monitor, it had a tail about 2.3 times the snout–vent length, it would have been about 7.9 meters long. Between the smallest and largest estimates is a difference in total length of about 51 percent of the shorter estimate. The greatest estimate of total length is still not as great as Lydekker's estimate of 1888, but it does show that his estimate does not necessarily imply a much larger body (snout–vent length). Nor was his estimate, as one might have feared from its great size, simply an unscientific guess.

Hecht (1975) pointed out, as we saw in the previous chapter, that *Megalania* is represented by relatively few caudal vertebrae. He also pointed out that caudals are the most commonly found bones of other fossil varanids and suggested that because they were rare for *Megalania* (as well as because larger varanids tend to have relatively short tails), *Megalania* also probably had a relatively short tail. If true, this suggests that estimates of total length of around 8 meters probably exaggerate the size of this beast.

Monitors may grow continuously through life (De Lisle 1996); however, since different species of monitors do show different characteristic sizes as adults, their growth is not unlimited. In young lizards, the joint (articular) surfaces at the ends of limb bones form elements separate from the shaft. These elements, known as *epiphyses,* allow the bone to grow at their junctions with the shaft while maintaining the form of the joint surface. This also happens in mammals, ourselves included; but in monitors the epiphyses do not completely fuse to the shafts as they do in mammals and other lizards (De Lisle 1996). We know this was also the case in *Megalania* because the femur of the

Alderbaran Creek specimen has a partially fused epiphysis at the joint for the hip (Fig. 3.7).

Continuous growth implies that, as with oras, mature individuals of *Megalania* at any one time and place would have been of several different sizes. Given that the largest known ora was about 3 meters long (Suzuki, Igarashi, and Hamada 1991) and the specimens measured by Auffenberg and his team averaged 1.7 meters long, or 57 percent of 3 meters, if *Megalania* had a similar size structure, then its average snout–vent length was about 2 meters. There is no evidence that, given a short tail (0.5 times SVL), the largest males reached more than 7 meters (total length), a little more than twice as long as the largest oras. However, no living varanids have tails proportionately so short; even in the ora the tail is longer than the snout–vent length (Auffenberg, 1981, Fig. 2-3). Admittedly, in large specimens (SVL of 1.25 meters), the tail is only slightly longer than the body. Assuming that adult *Megalania* had tails of only one-half the body length implies that Auffenberg's graph is applicable to *Megalania*—that is, that during its growth, the proportions of the body and tail changed in *Megalania* just as they do in *Varanus komodoensis*—and that Auffenberg's graph is not subject to the problems of extrapolation encountered in the following section. Both of these assumptions are likely wrong. Thus, we must at least consider the possibility that *Megalania* had a tail as long as the body and that the total length may have reached 7 meters or even a little more. Recorded lengths of oras are given in Table 11.

In oras, and in varanids in general, the males are usually larger than the females. So we would expect that the largest *Megalania* were males, probably reaching about 7 meters, a little more than twice as long as the largest male oras. Unexpectedly, this is the size Hecht (1975) estimated as the maximum, based on material that I think was incorrectly identified as *Megalania*. But he seems to have been right anyway: Nature is full of surprises.

Auffenberg points out that the sizes of oras are often overestimated, even by zoologists, and seemingly by about a third. So any Australians of 50,000 years ago who encountered a large *Megalania* would seem to see a very large and frightening beast indeed. If they were like modern observers, they would have guessed that a six-meter *Megalania* was eight meters long. Coincidentally, this is just a little short of how big Lydekker thought *Megalania* was.

Clearly, *Megalania* was the largest land-dwelling lizard known, although the marine mosasaurs were larger. No other carnivorous lizards approach this size, although some big herbivorous ones—for example, the Galapagos land iguanas (*Conolophus*)—reach a length of four feet (about 1.2 m). *Megalania* was also bigger than the ancient Mesozoic and Early Cenozoic monitors and monitor-like lizards. Crocodilians are another matter: Indopacific crocs (*Crocodylus porosus*) reach about seven meters in length—or at least they did before there were many people around. And the meat-eating theropod dinosaurs definitely included animals that were both longer and more massive than *Megalania*. But theropods, other than those that evolved into birds, have

Table 11

**Lengths of oras, from Auffenberg (1981) unless otherwise cited**

| Specimen(s) | Length (metres) | Length (feet) |
|---|---|---|
| largest male (Suzuki et al., 1991) | 3.09 | 10'1" |
| Senckenberg specimen (estimate: tail tip missing) | 3.04 | 9'8" |
| largest male measured by Auffenberg | 2.52 | |
| largest female measured by Auffenberg | 2.36 | |
| mean total length of 50 specimens from Komodo | 1.7 | c. 5'6" |

Note: The average total length of the 50 specimens from Komodo is taken from Auffenberg 1981, p. 23, where he also gives an average snout-vent length of 74.5 cm. This differs from the average body length of 55 cm. given by him on p. 152 (cited in chapter 5), presumably because the former excludes, and the latter includes, juveniles.

been gone for a long time, and *Megalania* was clearly larger than their replacements, the carnivorous mammals.

We should not, however, dismiss lizards too quickly: *Varanus sivalensis*, although very poorly known, was also big. The largest specimen, a broken humerus, was estimated by Lydekker (1886) to have been 2.4 cm in diameter (based on Plate 35: his statement in the text that the humerus was about 2.4 inches in diameter seems to be a typographical error). Lydekker estimated *V. sivalensis* as 11–12 feet (3.4–3.7 m) in length, probably by comparison with a skeleton of the water monitor *Varanus salvator* (although he does not explicitly say so). Data on the dimensions of ora skeletons are hard to come by, and the best I was able to do was a figured humerus (Surahya 1989, Fig. 70), approximately 1.5 cm in diameter. This does not allow much confidence in a comparison, but it does seem that *Varanus sivalensis* was about as big as an ora, maybe even a little larger.

*Megalania* was big, and we can estimate its length based on comparison with modern relatives. But, as we have seen in considering the proportions of the tail, these estimates vary depending on just which modern relatives we use. Oras are the current favorite because of their size and presumed close evolutionary relationship. But until a reasonably complete skeleton of *Megalania* is found and its proportions can be determined, we cannot be certain how big the largest individuals were.

## Estimating the Weight

If we are conservative in estimating length, we will probably not be far wrong in estimating the weights (masses, actually) of *Megalania*. There have been fewer estimates of mass than of length, although mass is a more important factor for understanding physiology, and hence life-

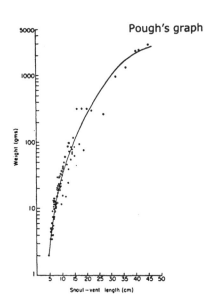

Pough's graph

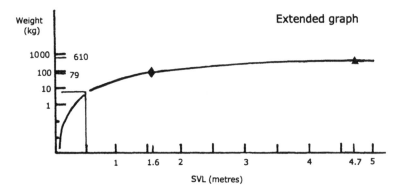

Extended graph

*Fig. 5.6. A reconstruction of Hecht's graph to estimate the weight of a 7-meter Megalania. Top, Pough's (1973) original graph. Bottom, estimated curve obtained by Hecht by adding Auffenberg's (1972) weight estimate of a 3.3-meter ora and Hecht's estimate of a weight of about 610 kg for a 7-meter Megalania.*

style. The published estimates vary from 227 to 2,200 kg, an even more drastic range than for estimates of length. The smaller weight is about that of an African lion, and the larger, 2,000 kg, is at the lower end of the range of weights for rhinos (*Rhinoceros unicornis*). Bob Bakker, in his 1986 book *The Dinosaur Heresies*, gives two estimates for *Megalania*, one of half a ton (about 454 kg) and another of 2 tons (about 1,814 kg). I suspect that the latter is a typographical error, since his estimate of the length (4.6 meters) is conservative, and the lower estimate of mass is more in keeping with it.

Only two people explained how they estimated the weights. Max Hecht used a graph published in 1973 by F. Harvey Pough, a herpetologist then at Cornell, to reach an estimate of 600–620 kg. Pough related the snout–vent length to weight for a variety of lizards, but all were less than a meter long. Even Bakker's estimated length is almost five times as large as the longest lizard measured by Pough, and Lydekker's almost ten times. Extrapolation to this extent is perhaps asking too much of

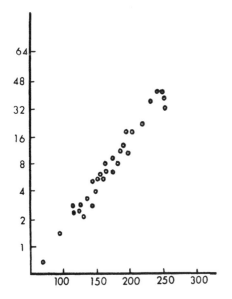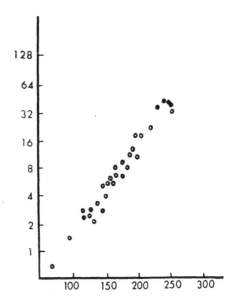

Fig. 5.7. Left, Auffenberg's graph of length vs. weight for the ora (Varanus komodoensis), as given in 1981. Note that the vertical scale is inconsistently calibrated. Right, the corrected version in which the vertical scale has been correctly calibrated, and a few points affected by this (all at the top) re-plotted.

Pough's graph (as we will see shortly), especially since the curve was fitted by eye. However, Hecht realized this, so he did not simply extend the curve Pough drew but augmented Pough's data with an estimate given by Walter Auffenberg of the University of Florida (1972) that a 3.3 meter ora would weigh about 79 kg (Fig. 5.6).

Auffenberg himself (1981) gave the largest estimate (for a *Megalania* 4.5 meters long): 2,200 kg. His estimate was based on extrapolating from the weights of oras of known lengths and assuming that *Megalania* had a similar process of growth. This is substantially bigger than any other estimates, except the second of Bakker (about 1,814 kg). However, there seems to be an error in the vertical scale of Auffenberg's graph (his Fig. 2-2); the logarithmic scale of the ordinate is incorrectly given. Correcting for this (Fig. 5.7), the estimate becomes about twice as big, around 4,400 kg. This is very much larger than any of the other estimates, which form two clusters around 500–660 kg and around 2,000 kg. The first cluster of estimated weights matches those of polar and grizzly bears (greater than the weights of lions or tigers), and the second is around those of rhinos and hippos. Among living mammals, only adult hippos and elephants match the corrected estimate of Auffenberg. The most massive living reptilian predator is the Indopacific croc (*Crocodylus porosus*), which in captivity reaches weights of 1,000 kg (Ross and Magnusson 1989). So at first glance, weights of 2,000–4,000 kg for a lizard seem extravagant. Various estimates of weight for *Megalania* are summarized in Table 12, and weights of some large mammals are given for comparison in Table 13.

The problem is that we are trying to work out the weight (mass) of a large animal from the relationship between weight and size for small animals. This relationship can be plotted, and the resulting plot forms a straight line. The difficulty is that almost any curve, if you look at a

Table 12
Estimates of the weight of *Megalania*

| Source | Weight (kg.) | Weight (lbs.) |
|---|---|---|
| Here, ca. 2.3 m long | 170–200 | 375–441 |
| Bakker (1986), 1st | 453.6 | 1,000 |
| Paul (1988) | 500 | 1,102 |
| Hecht (1975) | 600–620 | 1,323–1,367 |
| Bakker (1986), 2nd | 1,814.4 | 4,000 |
| Auffenberg (1981) | 2,200 | 4,850 |
| Auffenberg (1981), corrected | c. 4,400 | 9,700 |

sufficiently small segment of it, can look like a straight line. When we look at the curve for Pough's lizards, say 20 cm to about 1 meter long, we see a curve, apparently beginning to level off to become a straight line at the top. But presumably it can't level off (become horizontal), since we would expect a larger lizard to always weigh more than a smaller. We do not know how quickly the curve would approach being level when extended to include lizards 3 to 5 meters long. (To attempt to ascertain this, Hecht added Auffenberg's 1972 estimate of the weight of a large ora to Pough's data.) And since we do not know how the plot curves for these large sizes, we also do not know how far off estimates made from the plot might be.

In view of this problem, let's take another approach, recognizing that it is also uncertain. In 1985 J. Anderson, a zoologist at the University of Florida; A. Hall-Martin, a zoologist with the Kruger National Park (South Africa); and Dale Russell, a dinosaurian paleontologist then at the National Museum of Natural Sciences in Ottawa, published a paper giving a formula for estimating the weight of an animal from the size of its limb bones. They found that the sum of the circumferences of the humerus and femur (in mm) raised to the 2.73 power and multiplied by 0.078 gave the weight (in grams)—or $W = 0.078C_{h+f}^{2.73}$, where $W$ is the weight and $C_{h+f}$ the sum of the circumferences. If this is applied to the limb bones of a moderately large *Megalania* (i.e., about 1.5 meters in SVL) the circumferences come out to be 105 mm for the humerus (QM F865) and 137 mm for the femur (QM F4452). These give an estimate of about 250 kg. This is about the weight of an Indian tiger and is close to Bakker's lower estimate.

However, there is an uncomfortable risk here. The formula was derived for mammals and birds. These creatures have an upright stance, with the body weight supported entirely by their four (or two) limbs. In both fore and hind limbs there is one segment with but a single bone.

### Table 13
### Weights of mammalian carnivores
### (and others for comparison) (Walker 1975)

| Animal | Weight (kg.) | Weight (lbs.) |
|---|---|---|
| American black bear (*Euarctos americanus*) | 120–150 | 265–331 |
| African lion (*Panthera leo*) | 181–227 | 399–500 |
| Indian tiger (*Panthera tigris*) | 227–272 | 501–600 |
| Grizzly bear (*Ursus arctos*) | 360 | 794 |
| Alaskan brown bear (*Ursus middendorfi*) | 780 | ca. 1,720 |
| Polar bear (*Thalarctos maritimus*) | 320–720 | 705–1,587 |
| Northern sea lion (*Eumetopias jubata*) | 350–1,100 | 772–2,425 |
| One–horned rhinoceros (*Rhinoceros unicornis*) | 2,000–4,000 | 4,409–8,818 |
| Hippopotamus (*Hippopotamus amphibius*) | 3,000–4,500 | 6,614–9,920 |

This is the humerus in the upper foreleg and the femur in the thigh. The weakest parts of these bones should be those regions of the shafts with the minimal circumferences, and the measurements used here are these minima. Thus, for a quadruped, all of the body's weight must be supported by the two humeri and the two femora—all of it, of course, except that of the lower parts of the fore and hind limbs. The masses of these ends of legs is always quite small compared with that of the rest of the animal and so is ignored. For (standing) birds, their weight is supported by the two femora alone.

The smaller monitors, however, do not usually support their entire body weight on their four limbs. When walking, oras raise the belly and most or all of the tail off the ground as seen by several zoologists (Auffenberg 1981, 32 and Fig. 5-2; King and Green 1993, 43) and also as shown in videos. In this condition, all of their weight is carried only by the limbs, so the formula may be appropriate. After all, if they can do this, their leg bones must be strong enough to support their entire weight. If some of the body weight rests on the ground, this formula gives an estimate that is too low because it does not include the weight of that part of the body resting on the ground. Thus, it would underestimate the animal's mass.

Also, the formula was derived for animals with an erect or at least partly erect stance, something monitors do not have. It has been thought that in a sprawling stance the upper limb bones (humeri and femora) not only have to support the body's weight but also must resist being broken from the stresses inherent in the sprawling posture; so limb bones must be stronger for a sprawling posture than for an erect one. For me, at least, this made sense because of the experience of doing pushups. However, the human arm is adapted to supporting weight neither in an upright stance nor in a sprawling one—when it supports weight at all, it is carrying something, which puts tension, not compression, on the arm. Measurements of lizard limb bones reveal that they are no thicker (do not have a greater diameter) than in mammals of the same weight (Blob 2000). The limb bones of a large mammal, for example an elephant or a dinosaur or even a human, are held nearly vertical. If these bones are not strong enough, they will, as an engineer would say, fail in compression. In nonengineering terms this means they would be crushed. Small mammals—your dog or cat, for example—do not hold their limbs in as nearly vertical a position. Their limb bones would buckle (or fail in bending). Lizard limb bones, however, would do neither. Their greatest stress is encountered in twisting or torsion (Blob 2000), and it takes a thicker bone for any given length to resist torsion than bending or compression. As mentioned before, however, Blob found that lizard bones do not seem to be thicker. This implies that from the naïve analysis we might expect the formula of Anderson and his collaborators to overestimate the mass of *Megalania*, but the data of Blob do not bear this out. Thus, if anything, the formula would *under*estimate the mass.

But this is not the last of the problems, for there are no reasonably complete skeletons of *Megalania*. The humerus and femur used here did not even come from the same skeleton. In fact, they came from animals that may have differed in size (SVL) by about 25 percent (reckoned by comparing them to the lace monitor skeleton used for length estimates). This implies that for the lizard that contributed the larger element (the femur) this is an underestimate of the weight. Hecht estimated that the humerus we used derived from an animal with a snout–vent length of about 1.5–1.6 meters (my estimate is about 1.5 meters), thus giving it an overall length of 2.25–2.4 meters. Even if the tail were as long as the body, the overall length would still be about 3 meters, not substantially longer than a big ora.

This length may give us a way to see how realistic our weight estimate is. Does a 3-meter ora weigh around 250 kg? In 1972 Auffenberg estimated that a 3.3-meter ora (total length) weighed about 80 kg. His 1981 graph (corrected) suggests, instead, a weight of about 130 kg for a 3-meter ora. (Interestingly, the uncorrected graph suggests a weight of

Table 14
Table of estimated masses calculated or mentioned in text

| Method (ref.) | Taxon | SVL [TL] (m) | mass (kg) |
| --- | --- | --- | --- |
| Bakker 1986 -1 | *Megalania* | - | 454 |
| Bakker 1986 -2 | *Megalania* | - | 1,814 |
| Hecht 1975 | *Megalania* | 3.5 [7] | 600–620 |
| Auffenberg 1981 | *Megalania* (= ora) | 2.25 [4.5] | 2,200 |
| Auffenberg 1981 (corr'd) | *Megalania* (= ora) | 2.25 [4.5] | 4,400 |
| Anderson et al. 1985 | *Megalania* | 1.5 [2.25–3] | 250 |
| Auffenberg 1972 | ora | 1.65 [3.3] | 80 |
| Auffenberg 1981 (corr'd) | ora | 1.65 [3.3] | 130 |
| Blob 2000 (corr'd) | *Megalania* | 1.5 [3] | 80 |
| Blob 2000 (corr'd) | *Megalania* | 2.25 [5] | 320 |
| Hecht 1975 | *Megalania* | 1.5 [2.3] | 170–200 |
| Auffenberg 1981 (corr'd) | ora | 1.2 [2.4] | 32 |
| Auffenberg 1981 (corr'd) | ora | 1.15 [2.3] | 30 |
| Blob 2000 (corr'd) | *Megalania* | 2 [4] | 213 |
| Blob 2000 (corr'd) | *Megalania* | 2.85 [5.7] | 720 |
| Blob 2000 (corr'd) | *Megalania* | 3.8 [7.6] | 1,940 |

about 100 kg for a 3.3-meter lizard, not that far from his 1972 estimate of about 80 kg, suggesting that if the corrected graph is accurate, the 1972 estimate is too low.) The 130 and 80 kg estimates may not exactly agree, but they agree enough to suggest that 250 kg is an excessively large weight for an ora.

However, what seems to be the most reliable method for estimating the weight, and that usually used by biologists, is to work out mathematically the relationship between the size and weight from measurements of living animals. One takes a series of measurements, preferably as many as one can get, and from these derives a graph or an equation by the well-known method of regression. One can now buy computer programs that do this, thereby avoiding lots of tedious arithmetic and the possibility of making arithmetic errors. Fortuitously, such an equation appeared just in time to be included here. In the course of studying the effects of posture on the forms of the limb bones, Richard Blob at

Clemson College in Clemson, South Carolina, did just this (Blob 2000). Blob actually derived two equations for varanids, one that worked for the living species other than the ora, and the second that applied to all living species, including the ora.

If Blob's first equation does not apply to the ora, it seems unlikely that it would be appropriate for the even larger *Megalania*, so I shall use his second. Unfortunately, we discovered an error in the equation as printed in the original report, but Blob has provided the corrected version (pers. comm., 2002). This equation states that the logarithm of the mass of a monitor equals that of its snout–vent length multiplied by 3.435, from which is subtracted 6.009, or $\log(mass) = 3.435\log(SVL) - 6.009$. As with the formula of Anderson and colleagues, length is in millimeters and weight in grams. This is a simple equation: all one needs to do is look up logarithms, multiply, and subtract (or use a hand calculator). For a *Megalania* 1.5 meters long (SVL), we find a mass of about 80 kg—not very large, but not unreasonable either. An individual with a SVL of 2.25 meters (5 meters in total length) would weigh about 320 kg. (To help keep these and the following estimates clear, a summary of all of the mass estimates made in this section is given in Table 14).

Hecht's estimate of weight is presumably based on his largest estimated (total) length, 7 meters (which in turn is based on the claw that is probably not from a varanid at all). But to compare with our estimate from the circumferences of the limb bones, we have to scale down from Hecht's maximal length to estimate the weight of a smaller individual. Using a (total) length of 2.3 meters, intermediate between 2.25 and 2.4 (SVL of about 1.15 meters), we get a weight of about 170–200 kg, using Pough's graph (with the curve extrapolated to points at 3.3 meters and 79 kg and 7 meters and 610 kg, which are Hecht's values). This is much closer to our estimate of 250 kg using the method of Anderson and colleagues. But remember, this estimate was for a larger lizard with a snout–vent length of 1.5 meters, not 1.15 meters. For an individual 1.15 meters long (SVL) the weight should be nearer 200 kg. Unfortunately, without any limb bones from such an individual, its weight cannot be estimated by this method. Giving a range of values (170–200 kg), rather than a single value, arises from using Pough's graph without knowing exactly what his values were (they were not published), and therefore extrapolating his graph by eye rather than calculating the curve. Given all the uncertainties in our estimate of 250 kg, and the difference in sizes, these values are not a bad match. The estimate from Blob's revised equation for a 1.5-meter (SVL) lizard, 80 kg, is seriously smaller. However, because Blob's equation is based on closely related animals that also (presumably) live in much the same fashion, that equation seems the most reliable method of estimating the mass of *Megalania*.

In summary, for individuals of moderate size (for *Megalania*), Anderson and colleagues' formula gives 250 kg for an individual with a snout–vent length of about 1.5 meters, Pough's graph (extended with Hecht's values) gives about 170–200 kg, and Blob's (revised) equation gives about 80 kg. To give these estimates some perspective, remember that lions weigh around 180–225 kg, pumas about 100 kg, and American black bears (*Euarctos americanus*) about 120–150 kg. Auffenberg's (corrected) graph gives about 130 kg for an ora of the same length (1.5 meters in SVL), but this estimate requires extrapolating well beyond his data.

We expected that Anderson and colleagues' formula would overestimate the mass of *Megalania,* and in fact it seems that it does. Blob found that monitor limb bones increased more quickly in stoutness, as the animals increased in mass, than those of other lizards (he looked specifically at iguanas), so the limbs of large monitors should be stronger (and more robust) for their size than those of iguanas. Hecht noticed that the humerus of *Megalania* (QM F856) was much stouter than would be expected from comparison with that of the ora. This suggests that this trend of strong limb bones was continued in *Megalania.* And it supports our expectation.

We do not need to know precisely how heavy any individual *Megalania* was, just to have a reasonable estimate. After all, the weight of modern animals varies depending on how good the food supply is: when well fed and in good condition, they are heavier than when food is scarce; and they often vary in weight seasonally. Oras may vary in weight seasonally by 35 percent (Auffenberg 1981). We can even accept an estimate half or twice as big as the animal's weight, but we wish to avoid one that is five or ten times as great. The proportion of the tail length to that of the head and body probably would not much affect the results. After all, the distal part of the tail would be slender and hence weigh little: it might increase the estimates by 10 percent, but it would not double them.

However, people do have a fascination with the maximum weight and the greatest length—and not only for lizards. In oras, as generally among wild animals, few individuals reach the maximum size. For understanding the ecology of an animal, living or extinct, it is the average size that matters more than the rare individuals who grow to the full. That said, the Lau and Frost specimens (SVL estimated at 2.2 to 2.4 meters) seem to be close to the average, to judge from the sizes of the vertebrae held in the Queensland Museum. Blob's corrected equation gives a mass of about 320 kg for a 2-meter (SVL) individual. The largest specimen (QM F2942) gives an estimated body length of 3.8 meters, which in turn gives a mass of approximately 1,940 kg. The vertebrae in the collection suggest that lizards this size weren't unusual, although it is only to be expected that bias in preservation would favor the larger specimens. This estimate suggests that really large individuals of *Megalania* got as large as really large individuals of the Indopacific croc and were substantially more massive than most living mammalian predators, reaching weights close to those of polar bears.

## The Den

Many, or most, large monitors retreat to dens of some kind (Bennett 1998). Some make use of hollow trees or logs, or hide in thickets or crevices in rock. Many, however, construct or confiscate burrows (Fig. 5.8). Hollow trees or logs would seem inadequate for a mature *Mega-*

*Fig. 5.8. A goanna burrow in far southwestern Queensland (near Quilpie).*

*lania,* although appropriate for an immature individual. Thickets might be of service even to an adult, but crevices in rock would largely be restricted to the hilly and mountainous regions of Australia. Large areas of central New South Wales and central Queensland lack extensive outcrops of rock. In many such areas, the soil is deep enough to permit the excavation of burrows. Living wombats occupy some small parts of these regions and construct and maintain home burrows.

A burrow big enough to accommodate an adult *Megalania* would have been quite a feature. It is certainly possible that large individuals did not need a burrow since they would have been larger than any potential predators and, as we shall discuss in the next chapter, large enough to be protected from some of the consequences of extreme weather. Auffenberg (1981) discusses the burrows of oras in some detail but without explicitly presenting any diameters. His figure 5-12, however, indicates that some burrows were half a meter (or a little less) across. The burrows are also rather shorter than might be expected, averaging a meter and a half or less in length. Large oras doubled up inside with their tails lying alongside their heads and trunks. As Auffenberg points out, this posture is not universal among monitors; even immature oras sleep stretched out in their burrows. Thus *Megalania* may or may not have doubled up inside their burrows. The increased effort of excavating a large burrow—another example of the length-surface-volume relationship discussed later—as well as the increased risk of collapse suggest that burrows would have been as small as consistent with providing shelter.

Some idea of the diameter of a burrow could be given by the diameter of the rib cage of *Megalania*. Unfortunately, although several ribs are known, they are too incomplete for this purpose. Thus, we are left, basically, with guessing a diameter, bearing in mind the considerations of the preceding paragraph. From the difference in size between *Megalania* and an ora, between half a meter and a meter would seem a reasonable estimate for the diameter, and two to three meters for the length.

There may have been another large burrowing creature in Australia contemporaneous with *Megalania*—the giant wombat *Phascolonus*. The preparation staff at the Queensland Museum worked out—by simply increasing the diameter of a modern wombat burrow by 50 percent to match in the increase in size of the giant wombat over those of the living ones—that if *Phascolonus* dug burrows, they would have been about 35–40 cm in diameter. How long their burrows might have been is a more difficult question, but presumably several meters is not out of the question.

In fact, we have no evidence that either creature burrowed. This possibility is based only on the circumstance that all living wombats burrow, as do many living monitors. If the giant wombat did construct burrows, these may have been suitable for small megalanians as well. On the other hand, the effort of excavating such large burrows and the risk of their collapse suggest that both creatures may have relied on other types of shelter, at least as adults.

## Hunting

There is no useful evidence regarding how *Megalania* went about its hunting. Still, it must have had some strategy. Auffenberg observed that oras lay in wait, hidden, near game trails and then rushed out and attacked their meal-to-be. Animals using this strategy are known as *ambush predators*. However, French and joint Japanese-Canadian videos show that at least sometimes oras act like *pursuit predators*—those, like wolves and cheetahs, that actively run down their prey. To visualize an ora galloping down a deer or boar certainly requires a robust imagination, but each program includes one sequence, 13 and 6+ seconds long, respectively, of large oras rapidly chasing prey—in the first case, a kid, and in the second, a deer. The sequences were long enough to show that these were not just quick lunges—in other sequences the lunges took about five seconds—but they did not show the entire chases. The shorter sequence is included because the ora twice attacked the deer by lunging during the sequence. Interestingly, Auffenberg suggests that oras may be able to track prey for some distance by following scent trails. *Megalania* may well have been an ambush predator, but may also have tracked slow or already-bitten prey.

The ambush strategy of oras has another aspect, the use of septic bacteria in subduing prey. Meat eaters generally have foul-smelling breath because their mouths host bacteria living on scraps of meat caught on their teeth and gums. Oras benefit from this in that particularly virulent infectious bacteria inhabit their mouths. When attacking prey, these bacteria may infect the wounds and seriously weaken or even kill the animal (Auffenberg 1981). The deer chased by the ora in the video sequence mentioned above had previously been bitten and was presumably weakened by infection. Both monitor and bacteria benefit: the ora by gaining food more easily (expending less effort), and the bacteria by acquiring the resources to support a (temporarily) larger population. This seems to be an adaptation on the part of the bacteria because increased virulence will directly benefit those bacteria that exhibit it. On the other hand, because the prey animal is not immediately struck down but can escape and may collapse elsewhere, it may be found by some lucky ora other than the one that bit it. So the benefit to the oras is less immediate than that to the bacteria—unless, of course, the ora can track the bitten prey so that the benefit usually accrues to the same individual that has invested effort in (i.e., bitten) the prey. Although there is no indication that *Megalania* used this tactic, it is quite plausible that similar bacteria—perhaps even the same—lived in their mouths, and these could well have infected prey animals. Incidentally, there is no evidence that the bacteria infect individuals bitten by other oras during fights among themselves: they seem to be immune (Auffenberg 1981, 154).

Large monitors seem to be, in a sense (and with only a wee bit of exaggeration), creatures of routine. Auffenberg found that oras habitually followed similar routes through their territories when they were out foraging. Presumably, they were checking places where they were likely to find food. Some of these routes are so well used that distinct paths are formed—although the oras apparently followed scent tracks, not the "foot paths" on the ground. Frank Seebacher, a physiologist now at the University of Sydney, told me of similar behavior among Australian lace monitors. If this is characteristic of large monitors, we would expect *Megalania* to have done the same—although its pathways have now long been obliterated.

Neither the size of *Megalania,* nor its strong teeth and claws, nor even its quite possibly toxic bite would have been the key feature that made it a ferocious hunter. Instead, that would have been its physiology. In 1964 George Bartholomew, then at UCLA, and Vance Tucker,

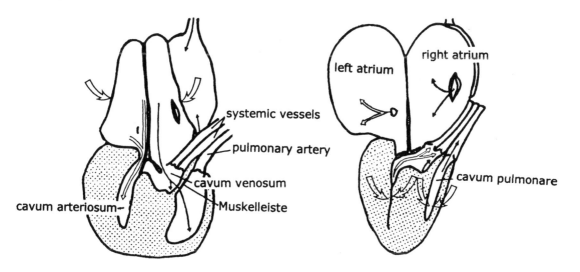

then at the University of Queensland, published the results of the first general physiological study of monitors. They found that monitors had a higher-than-expected metabolic rate, greater than in other lizards. This was later confirmed specifically for oras in the wild, compared with iguanas (Green et al. 1991). The metabolic rate of a varanid could be increased during activity until it matched or surpassed the basal rates of mammals. It was these kinds of results that led Auffenberg, early in his book on oras, to warn his readers that expectations derived from other lizards could be seriously misleading in regard to varanids. Other work on monitors (Bennett 1973) showed that they could sustain higher levels of activity because their blood did not lose its capacity to transport oxygen during activity as quickly as did that of other lizards. The implication is that monitors could be more active and tire less quickly than other lizards of the same size.

The hearts of monitors are also different from those of other lizards. Lizards' hearts do not have completely separate ventricles as found in those of birds and mammals. As a result, there would seem to be the opportunity for oxygenated and deoxygenated blood to mix in the flow to the body, but physiological experiments indicate that this usually does not occur. In varanids, the blood flow is more like that of birds or mammals (Millard and Johansen 1974) than like that of other lizards, in that the blood flowing to the body (systemic circulation) is supplied under greater pressure than that flowing to the lungs (Fig. 5.9). The heart of oras is constructed to further support this, although to my knowledge no one has verified this effect in living animals. Furthermore, the mechanics of breathing in monitors is also unusual, but to understand this requires a short digression on breathing of other lizards.

In 1987 D. Carrier, of the University of Utah in Salt Lake City proposed that lizards are subject to a "trade-off" in function between moving about and breathing. In simpler language, this means that the lateral flexing of the body that accompanies walking or running in

Fig. 5.9. Blood flow in a varanid heart (based on Varanus bengalensis and V. gouldii). Left, blood flows from the lungs into the left atrium and returns from the body into the right atrium. Oxygenated blood flow is indicated by the narrow white arrows, and flow of deoxygenated blood by the black arrows. Then the atria contract (indicated by the broad white arrows), and blood flows into the cava (right). Oxygenated blood flows into the cavum arteriosum, and, when the cava contract (broad white arrows), into the systemic vessels to the body. Deoxygenated blood flows from the right atrium into the cavum venosum, across the Muskelleiste (an incomplete partition between the cavum venosum and the cavum pulmonare) and into the cavum pulmonare. When that cavum contracts, the deoxygenated blood flows through the pulmonary artery to the lungs. In spite of the incomplete partition (Muskelleiste) between the chambers containing oxygenated and deoxygenated blood, there is only inconsequential mixing of the two bloods. (From Webb et al. 1971.)

Fig. 5.10. Oblique dorsal view of an iguanid lizard walking, showing the lateral flexion of the body (and head).

lizards (Fig. 5.10) interferes with the efficiency of breathing. Specifically, flexing prevents effective filling and emptying of the lungs when running because lizards typically breathe by drawing air into both lungs together. The air is drawn in by rotating the ribs on the vertebral column, which thereby increases the volume of the rib cage and hence fills the lungs. When the vertebral column is flexed to one side, the volume of the rib cage on that side is reduced, while on the other it increases. This occurs when the back muscles involved act unilaterally, as they do during walking and running. Thus, the stamina of the animal is less than it would be if some different method of drawing air into the lungs were used. This is a somewhat imaginative exercise, for there is no other method of pulling air into the lungs for most lizards. It turns out, however, that varanids have developed a modification of their method of breathing that overcomes the disadvantage (Owerkowicz et al. 1999). Instead of drawing in air by suction (the result of increasing the volume of the rib cage), they force air into their lungs by "swallowing" it (Fig. 5.11). It is not exactly swallowing, however, since they use the windpipe, or trachea, not the throat (esophagus), in a process known to physiologists as "gular pumping." This can be—and is—done while running to overcome the trade-off and provide more stamina. Thus, monitors can more effectively supply oxygen to their lungs during exercise and more effectively carry it to their muscles than other lizards can.

A study of monitor diets by Jonathan Losos and Harry Greene (1988) of the University of California at Berkeley reported what is quite

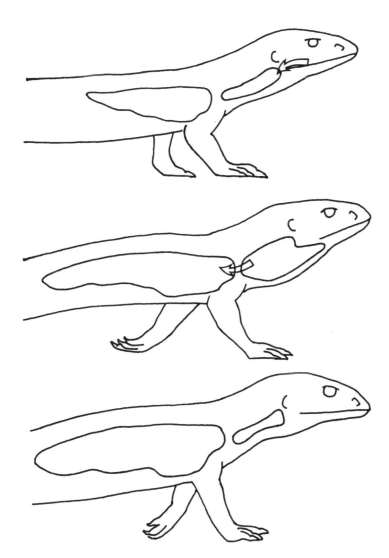

Fig. 5.11. Gular pumping in a varanid. Top: the lizard first forces air from the mouth cavity into the throat (trachea). Middle: having filled the trachea, the lizard then forces the air into the lungs (bottom). Air is forced into the lungs because the pressure is higher in the trachea, rather than being sucked into the lungs because the pressure in the lungs is lower, as during typical breathing. (From Owerkowicz et al. 1999.)

possibly an effect of this physiological development. They noted that varanids were more effective at seizing fast-moving prey animals than related (anguimorph) lizards. This is, presumably, an indication of enhanced prey-catching ability in monitors.

These physiological features are all found in modern varanids and hence can reasonably be expected in the more recent of the fossil ones as well. This implies that *Megalania* was not only big but that it was capable of strenuous activity in catching prey as well as having stamina more like that of a placental mammal than of a lizard. In other words, as we implied at the beginning of this discussion, it came equipped with the physiology to be a ferocious hunter.

There is one more physiological feature of monitors to mention: their ability to deal with temperatures above or below those they find optimal. Animals (and other objects) heat or cool in response to the temperature of their surroundings at a rate that depends on their surface area. For a doubling in length, the surface area will increase by four times. So we might think that for two individuals, one twice the length of the other, the longer individual will heat up (if the environment is warm) four times faster. In fact, just the opposite is true. The volume of the larger is eight times that of the smaller, and it is this volume that must be warmed to heat the animal. So the larger creature will heat less rapidly when the weather is hot, and retain heat longer when it is cold, than the smaller (Fig. 5.12). This effect is known as "thermal inertia." Together with the appropriate behaviors, such as basking in the sun to heat and resting in the shade—or in pools of water—to cool, a large animal can effectively control its body temperature. This can be seen in some living creatures. For example, large leatherback turtles can maintain a body temperature of up to 18°C above that of the water in which they swim, both at sea (Paladino, O'Connor, and Spotila 1990) and under laboratory conditions (Frair, Ackman, and Mrosovsky 1972). Large individuals of the Indopacific crocodile can also do this (Seebacher, Grigg, and Beard 1999).

Auffenberg found that oras were more independent of their environmental temperature than smaller monitors, and this would also have been the case for *Megalania,* which was even larger. Thus, we can reasonably expect that *Megalania* could hunt in cooler weather than could living monitors, even if only a little more, which is appropriate for the generally lower temperatures of the Pleistocene. On the other hand, we can expect that in prolonged cool periods, *Megalania* would cool off and would not need to hunt as much as when the weather was warm. Mammals, of course, being endothermic (i.e., generating their own body heat from metabolic processes) would need to eat, and hence hunt, more when the weather was cooler.

Oras tend to rest at places that are close to regions of both high and low temperature (near steep thermal gradients). For example, their resting sites would be at the margins of forests, close to both the heat of the grassland and the cool of the woodland. This is presumably so they can take advantage of either place—to warm up when they are cool or cool off when warm—without having to walk far. This tactic is also appropriate for an ambush hunter, since woodland margins often have denser vegetation than either open plains or forest interior and thus provide cover for large, hungry lizards. However, for reasons discussed in the next chapter, large individuals of *Megalania* may have taken more than a day (i.e., 24 hours) to significantly warm or cool. So although they may have lurked at woodland margins to take advantage of the cover, this was probably not to be near sources of heat or places to cool off.

We have built a reasonable picture of how *Megalania* hunted and why it was able to be a successful hunter, but did it actually do this, or has something been overlooked? In other words, how do we test these ideas? Accepting that *Megalania* was not a pursuit predator in the sense that cheetahs and dogs are, there is no way to tell from its skeleton whether it might have been an ambush predator or whether it chased prey over short distances. After all, oras seem to do both without obviously reflecting these behaviors in their skeletal features. Did *Mega-*

Fig. 5.12. The greater the volume (i.e., the larger) the lizard, the more slowly it will lose (or gain) heat. Here the larger is twice as long as the smaller, thus has eight times the volume, and so loses one-eighth as much heat in any given unit of time. The amount of heat lost is shown here as the area of the arrows.

*lania* subdue prey with bacterial infection? Again, we can't be certain: fossil kangaroo bones do show signs of infection, but there are other ways to get infected than by being attacked by *Megalania,* and an animal infected by the oral bacteria may not have escaped being eaten long enough for the infection to have affected the bones. What we have is a "best guess" scenario—but it is still a guess.

## Feeding

With some idea as to how *Megalania* hunted, let's now look at what happened when they had acquired their meal: how they may have fed. Unlike other large tetrapods, varanids have flexible joints at certain positions in their skulls, a condition called "cranial kinesis." This is not the case in mammals (or crocodiles), whose skull bones are sutured together to form a single rigid structure. The only time there is movement between the bones of a mammalian skull is at birth, when the baby's skull may flex to successfully pass through the birth canal. Otherwise motion is permitted only at the joints for the jaws, and this motion is adjacent to, not within, the skull. For large animals, it is generally thought that flexible joints (kinesis) would weaken the skull. This is not only the case for crocodilians and large carnivorous mammals but even for large predatory dinosaurs such as *Tyrannosaurus,* although some large dinosaurs—*Allosaurus* for example—did have kinetic skulls.

Max Hecht thought the skull of *Megalania* was sufficiently large that the flexible joints would have been lost. At the time he wrote this, before 1975, rather little of the skull of *Megalania* had been found. Paleontology is not usually regarded as a predictive science like physics; more often it is thought to be purely descriptive, like astronomy. Nonetheless, here is a case where Hecht's understanding of functional anatomy led him to make a prediction. This prediction could not be tested until fifteen years later when Ian Sobbe found further bones of the skull roof. These were the frontal and parietal, and the joint between these is flexible in other monitors—but not in *Megalania.* Max was right: here the joint is firmly interlocking. Apparently the skull of such a large monitor would have been significantly weakened by having flexible joints. (So why the skull of *Allosaurus* apparently was not weakened is yet to be explained, but it may have something to do with having had different hunting tactics.)

Teeth are usually fairly accurate indicators of diet. Eating is obviously important to animals—just consult your pet dog or cat on this—and what can be eaten and how depends on the teeth. Deer and dogs, for example, have distinctly different kinds of teeth, each suited for their diet. So it is generally accepted among paleontologists that some idea of what an animal ate can be gained from examining its teeth. However, inferring things about the past is not so simple. My major professor at UCLA, E. C. Olson, used to tell about skates in Tomales Bay (California). These fish had massive crushing plates of teeth, and one could no doubt work out just how thick a clam or mussel shell they could break. However, these skates didn't break clam or mussel shells at all; they ate worms (annelids). Carnivorous mammals have teeth that are well suited for cutting meat, but as your dog or cat can tell you, now and again they fancy a bit of grass. I once had a dog who enjoyed sauerkraut and another who picked peaches (but only very low ones) off trees and ate them (both were German shepherds). The form of the teeth tells what an animal *can* eat, and that form probably evolved because that is what they usually ate, but animals do on occasion choose to eat other foods. So the teeth of *Megalania* tell us what it *could* eat, but don't exclude the odd nibble of leaves or something else equally unexpected.

In *Megalania* the teeth are sharp, flattened, wickedly curved blades (Fig. 5.13). They are obviously well suited for cutting and slicing: the teeth of a carnivore. They are similar to, but rather larger than, the teeth of an ora, teeth that can be observed put to use cutting meat. Such teeth are also found in carnivorous dinosaurs, but not among meat-eating mammals (except for canines in some—extinct—cats). The edges of the tooth are not smooth but bear minute cusp-like structures, or serrations. The teeth are strongly pleated or wrinkled at the base, with grooves extending about a third of the way up the crown that perhaps retained bits of meat to support bacterial colonies. At its base, each tooth forms a rounded, flattened plate that attaches to the inside of the jaw. Most lizards do not have teeth set into sockets as mammals and crocodilians do; their teeth are either fused to the edges of the jaws or, as in monitors, attached by connective tissue to the medial sides of the jaws (Fig. 5.14). (This is the pleurodont condition mentioned in the previous chapter.) In such lizards the connective tissue periodically loosens, the bases of the teeth are resorbed, and the teeth are shed to be replaced by new ones.

*Megalania* teeth have a rounded fore edge and a sharp after edge, this latter bearing most of the serrations, with only the upper third of the fore edge being serrate. Like the serrations of a steak knife, these facilitated the slicing of meat. William Abler (1992), in Chicago, conducted experiments designed to understand the function of carnivorous dinosaur teeth (specifically those of tyrannosaurs). Certain of his conclusions, those pertaining to features of tyrannosaur teeth that are also found in *Megalania* teeth, are relevant here. Abler found that teeth with large serrations required the exertion of more force along the length of the tooth (what he called "drawing force") but less force directed perpendicularly against the meat ("downward force") to make cuts

Fig. 5.13. (top) Close-up of a Megalania *tooth*. *The recurved, blade-like form is obvious, and steak-knife-like serrations can be seen along both edges. This tooth was found in Pleistocene deposits near Burketown, far north-western Queensland. This tooth was broken, so although the basal striae (plicidentine) can be seen, it is not as clear as in a complete tooth. (Courtesy of M. Archer.)*

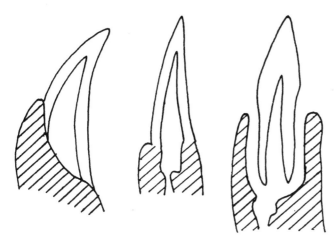

Fig. 5.14. (bottom) Modes of *tooth attachment. Left, in pleurodont attachment the tooth attaches to the inside of the jaw. This is found in many lizards, including varanids. Middle, in acrodont attachment the base of the tooth is fused to the jaw bone. This form is also found in many lizards, including the agamids. Right, in thecodont attachment, the root of the tooth is set into a socket in the jaw bone. This form is not found in lizards, but it occurs both in mammals and in archosaurs, such as dinosaurs and crocodilians.*

equal to those of teeth with smaller serrations. A model tooth without serrations required greater force of both kinds. So although the optimal size of the serrations depends upon how much force is exerted in biting (drawing force) and how much from pulling the head back from the prey (downward force), any serrations are better than none. And if most of the force is exerted in pulling back on the prey, as oras can be seen to do in some videos, small serrations are more effective than large. The teeth of *Megalania* (like those of tyrannosaurs) have relatively small serrations. Abler found that meat fibers caught between the individual serrations were extremely difficult to remove, which makes certain their retention in the mouth, where they provide homes for the septic bacteria. The serrations cut fibers (he used nylon monofilaments) with unexpected ease (needing only 5 percent of the tensile strength of the fiber), even when the serrations were worn. This may well be helpful in cutting hide.

Jim Farlow, a dinosaurian paleontologist at the joint campus of Indiana University–Purdue University Fort Wayne, and his colleagues (1991) studied the relation of the length of the tooth at its base to the height of the crown—in other words, the proportions of the tooth when seen from the side. They found that the proportions were reasonably constant for a variety of meat eaters, including varanids, theropod dinosaurs, toothed birds, and even some mammalian canines. (But modern crocs, and most of those contemporaneous with *Megalania,* probably have different proportions.) So it is plausible to think that these teeth all acted in much the same fashion for much the same purpose.

Our best guess is that the obvious is correct: *Megalania* teeth are suited for cutting meat. Although there are two monitors, Gray's monitor (*Varanus olivaceus*) and Panay monitor (*Varanus mabitang*) in the Philippines, that eat some fruit (in addition to invertebrates and birds), the rest specialize in meat eating, and we have no reason to doubt that *Megalania* did as well. Oras avoid ingesting plant material to the extent that they vigorously shake the stomach and intestines of their prey to scatter the vegetable contents before swallowing these organs. So we may suspect that *Megalania,* like oras but unlike cats and dogs, probably did not nibble plants.

A notable feature of monitors is the efficiency of their eating (Auffenberg 1981, 209). There are two aspects to this feature, first how much of the carcass is consumed, the inverse of what Auffenberg termed the "waste factor." Mammalian carnivores—Auffenberg looked at lions, leopards, pumas, tigers, wolves, and African wild dogs—usually leave behind about 30–50 percent of the carcass, but oras usually left behind only about 10 percent (Fig. 5.15). Presumably, the more that is ingested, the more that is digested and can thus provide the lizard with energy, which leads us to the second aspect, how much of the energy content of the ingested food is actually utilized, or assimilated, by the predator. Australian zoologists Dennis King and Brian Green, in their 1993 book on Rosenberg's goanna, state that monitors (and predators in general) assimilate 80–90 percent of the energy content of prey animals. Thus, oras should consume about 90 percent of the carcass and 80–90 percent of the consumed energy content, utilizing about 70–80 percent of the energy content of the carcass. Because of the lower waste factor, monitors should be (and have been) more efficient at gaining energy from their prey than other competing lizards and mammals. Auffenberg suggested that this accounts for the relative lack of scavengers on the islands inhabited by oras. This may well have been a characteristic of *Megalania* and may have given it a

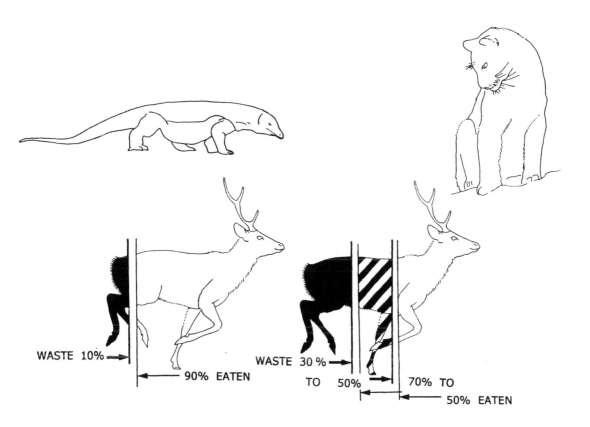

WASTE 10% → ← 90% EATEN

WASTE 30% → 70% TO
TO 50% 50% EATEN

Fig. 5.15. Waste factor in oras
and carnivorous mammals (here
represented by a puma). Oras
usually devour about 90 percent
of the prey (here arbitrarily
represented as a deer), and leave
behind ("waste") about 10
percent. Mammalian carnivores
are more fastidious, and consume
50 to 70 percent of the carcass,
leaving behind 30–50 percent.
(Based on data from Auffenberg
1981.)

competitive edge over the Australian marsupial meat eaters. Unfortunately, the waste factors of carnivorous marsupials are not known, either because they are extinct or—for the few survivors, like Tasmanian "devils"—because they are rare and protected.

So what did *Megalania* eat? Judging from observations on living monitors, anything they could catch. This response is less of a joke than it may seem at first (and Auffenberg said much the same thing of oras). The Queensland Museum has a mummified specimen of a perentie (*Varanus giganteus*) that attempted, quite unsuccessfully, to eat an echidna (Fig. 5.16). The monitor died in the attempt, and both carcasses dried out. The lizard may have tried this at a time when no other prey was available or scavenged a dead echidna. This points up that monitors are able, or at least try, to eat items other than their usual food when that is scarce. They may "regress" and rely on food that they preferred when they were much younger, or they may try items not usually eaten—like (presumably) echidnas.

There was no shortage of potential prey animals for *Megalania*: diprotodonts, kangaroos (both giant and "normal"), giant wombats, and large flightless birds (dromornithids) were all contemporaries. Large prey is suggested not only by the size of *Megalania* but also by the contention that both average and maximum prey size are correlated with snout–vent length in living monitors (Pianka 1994). Unfortunately, no further details are given. Toothmarks of *Megalania* have been

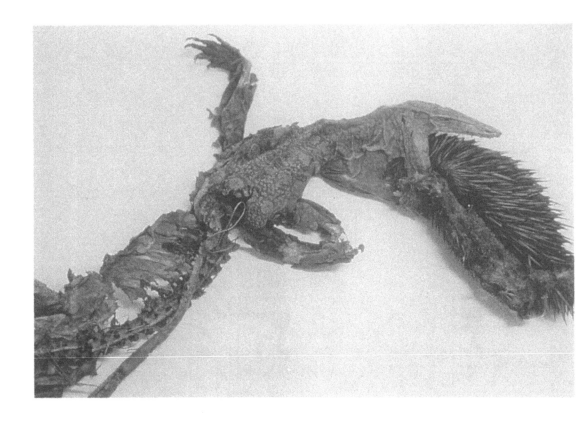

Fig. 5.16. A favorite exhibit in the Queensland Museum. The mummified carcass of a perentie (Varanus giganteus) that died attempting to eat an echidna (Tachyglossus aculeatus). This presumably occurred during a drought or other time of food scarcity, since even a goanna wouldn't usually consider trying to eat an echidna.

suggested as an explanation for the scratches found on some fossil bones, but this has yet to be verified. The only possible indication of gut contents is a single wombat incisor found with the specimen from Alderbaran Creek. Since several ribs of this specimen were found, the rib cage presumably was there and gut contents could have been preserved. Furthermore, the surface of this tooth seems to have been etched, quite possibly by the stomach acids of the *Megalania*. So at least one *Megalania* may have eaten at least one wombat.

Around the end of the nineteenth century, the large African carnivorous mammals varied in the regard of Europeans, depending on whether they were scavengers or predators—or more accurately, whether they were *thought* to be scavengers or predators. This view has suffered a kind of resurrection in similar discussions about the eating habits of theropod dinosaurs, especially *Tyrannosaurus*. Thanks to extensive and careful observation, we now know that lions, long thought to be pure predators, do scavenge carcasses and that hyenas, long thought to be mere scavengers, do kill their own prey. The boundary between predation and scavenging has become blurred. Such distinctions are much more important to people than to the animals: the animal's chief concern is with getting fed, preferably with as little effort as possible. This is quite understandable when you might not know where your next meal is coming from. So if a meat eater can obtain a fresh carcass (or in some cases, at least an edible carcass), this would not be looked upon as degrading, but more as a gift from the patron deity of predators: food without the effort of catching and killing it. Varanids do not hesitate to scavenge (but oras do stop at mummified carcasses). *Megalania* presumably did the same, being an active hunter much of the time, but not scorning a fresh carcass when one was available.

## Population Biology

Before you can eat your prey, or even find it, it has to exist. Meat-eating animals—whether lizards or lions—have to live where adequate food (prey) is available. This is why big predators are one thing you do not have to worry about in deserts, be they the sands of the Sahara or the snows of the Antarctic. The situation, however, is more subtle than that. Places that may appear to us to be able to support large predators, like jungles or rain forests, often actually cannot because there are inadequate numbers of prey animals. This is why lions live on the savannas of eastern and southern Africa, not in the rain forests of central Africa.

The key item here is what ecologists call the predator-prey ratio, the proportion of prey animals to each meat eater. This is not a constant, but depends on a number of factors, such as the size of the predator. However, the key factor is the metabolic rate of the carnivore. The metabolic rate represents, in effect, the cost of being alive. Mammalian predators, lions for example, have higher metabolic rates than reptilian predators such as crocs. Thus, lions need to hunt regularly, but some crocodiles reportedly can get by on one good feed a year, that is, when the wildebeest and zebra cross—or attempt to cross—the Grumeti River of the Serengeti (Tanzania) in which they live. The average mammalian predator requires about ten times as much food in any given period of time as the average reptilian predator of equal weight because of the greater mammalian metabolic rate. Monitors are not your average reptilian predators, but they can still get by on much less in a given amount of time than can, say, a cat of equal mass.

Auffenberg estimated the predator-prey ratio for oras on Komodo at 1 adult per 105 prey animals. This is for big oras of 2.5 meters long or more. He also presents estimates for some big cats, 1 puma per 200–360 prey animals, for example; or 1 lion per 260–360 prey, and 1 tiger per (about) 350 prey. We can't work out predator-prey ratios for *Megalania*, if only because we don't really know the abundance of its prey animals (which we can only guess in any case). But perhaps we can at least try to work out how abundant *Megalania* was.

Auffenberg estimated the biomass for oras both on the island of Komodo and on the nearby island of Padar. Biomass, in this case, is the total of the masses of each ora for a given area, here a square kilometer. Notice that biomass is not a property of the oras as individuals but is a property of the population of oras. The average body length of the oras of Komodo was about half a meter (55 cm, to be exact), and the biomass was estimated as about 33.6 kilos per square kilometer. On Padar the average size was about the same, but the monitors were not as common, and the estimated biomass was about 23 kilos per square kilometer, smaller by about one-third. We have an estimate of average mass for *Megalania*, 320 kg (for SVL = 2.25 m).

Equations relating the number of individuals in a given area (the density of individuals per square kilometer) to mass (in kilograms) are given by Jim Farlow (1993). He found that density (individuals per square kilometer) equaled 66.07 times mass (in kilos) raised to the −1.01 power, or $d = 66.07m^{-1.01}$. For a 320 kg *Megalania*, this gives a range of about 5 square kilometers. Another estimate can be made based on the work of Frederick Turner and his colleagues (1969), then at UCLA, whose research related body weight to the area of the home range for lizards. They found that the area of the home range was equal to 171.4 times the body weight to the 0.95 power. Expressed as an equation, this is $A = 171.4W^{0.95}$, where A is the area of the range in meters squared, and W is the weight in grams. For our "average" *Megalania*, this gives a range of about 29 square kilometers. This is not an exact match with the estimate from Farlow's formula, being about six times as much, but—as physicists would say—it is within the same order of magnitude. The "agreement" could be much worse. For comparison, lions can have ranges of 4,000+ square kilometers, although this is admittedly at the upper end of the spectrum of range sizes. The average is around 200 (Schaller 1972).

The estimate from the formula of Turner and colleagues may overestimate the range. However, Auffenberg found that the formula of Turner and his colleagues *under*estimated the size of the range for oras, giving only about 60 percent of the sizes he found. Phillips (1995) found that this formula also underestimated the range of the white-throated savanna monitor (*Varanus albigularis*) in Namibia, predicting about 3 percent of the actual range. He also noted that it underestimated the range of the grey monitor (*Varanus griseus*) in Israel, although not by so spectacular an amount (predicting about 12 percent of the range). It may be significant that Turner and his co-workers did not use data from monitors in deriving their equation. But here, if anything, the formula overestimates the range for *Megalania*, and given the uncertainties in estimating the mass and extrapolating the formula to a creature the size of *Megalania*, we really have no cause to complain about the results.

So are these estimates acceptable? Probably. A range for *Megalania* $1/10$ or so the size (area) of that for a lion fits in with the average food requirements of an ectothermic lizard, being about 10 percent those of an endothermic mammal. But this assumes that the animals have at least approximately the same mass, and our estimated mass for *Megalania* is about half again that a lion. A *Megalania* of that mass would require a little more than 10 percent (presumably about 15 percent) as much food (per unit of time). All other things being equal (such as ease of prey capture), such considerations suggest that the result from Farlow's formula may be low, but not substantially so.

These estimates indicate that *Megalania* would have had ranges of 10 to perhaps 40 square kilometers, larger if there was substantial overlap of ranges. Hence, individuals may have been fairly rare on the ground.

The only information we have with which to independently assess these estimates is how common the fossilized bones and teeth of *Megalania* are. As already mentioned, the fossils are, in fact, quite rare. Although this does not imply that our estimates are correct, the estimates are at least consistent with the rarity of the fossils. So the ranges of *Megalania* are unlikely to have been much below a kilometer square or much over, say, 50 square kilometers. Of course, the ranges presumably varied with the abundance of prey animals. There is no indication of more than a single individual of *Megalania* at any one site, and often only a single bone or tooth was found per site, which is again consistent with large ranges.

We can get a better notion, although by no means exact, of how abundant *Megalania* was by comparing the number of their teeth found to those of crocodilians (Willis and Molnar 1997b). Fossils of *Megalania* are found in the same deposits as those of the large crocodilian *Pallimnarchus*. Both kinds of animals produce, and shed, teeth throughout their lives. The Queensland Museum collections hold isolated fossil teeth of both kinds. There are about eighty times as many crocodilian teeth as *Megalania* teeth, in spite of the fact that both professional and amateur paleontologists collecting for the museum have generally been searching for *Megalania* teeth more enthusiastically than for crocodilian teeth (although crocodilian fossils are also rare on the eastern Darling Downs, so croc teeth have been specifically sought for). In a lifetime, a single modern crocodilian produces more than 350 teeth (Edmund 1962; 1969). Let us assume that *Pallimnarchus* did also. (There actually were some Indopacific crocs around at this time as well; however, since our rate assumed for *Pallimnarchus* is based on the rate of tooth production in modern crocs, this should not introduce any discrepancy.) If *Pallimnarchus* lived for ten years, this is an average of 35 teeth per year; if for longer, less per year. (This age of ten is chosen arbitrarily to simplify calculation. The Indopacific croc almost certainly lives much longer than ten years, as, probably, did *Pallimnarchus*. However, using a "short" age overestimates the amount of teeth *Pallimnarchus* produced per year and hence reduces the risk of underestimating the abundance of *Megalania* in this exercise.) Oras produce 200–250 teeth per year (Auffenberg 1981). If we assume that *Megalania* did likewise, we see that unless somehow the process of fossilization discriminated against *Megalania* teeth and in favor of crocodilian teeth, there must have been many more crocodilians than giant monitors. If all shed teeth of one *Pallimnarchus* and one *Megalania* were collected, the ratio would be at least (approximately) one croc tooth to five monitor teeth; if *Pallimnarchus* lived as long as Indopacific crocs, it would be much less than one to five. Instead of one to five, it is actually eighty to one. So if there were equal numbers of individuals of both kinds, 400 monitor teeth would have been lost for every one preserved, or there were approximately 400 crocs to every *Megalania*. This was apparently a rare animal indeed.

Paleontological issues are seldom as simple as has been made out here. Crocodilians lived in and near watercourses, where their fossils (and those of *Megalania*) were deposited, but *Megalania* presumably did not. Maybe this could account for the difference, but then again maybe not. Mammalian teeth are also much more common in these deposits than those of *Megalania*, and mammals produce fewer teeth per lifetime than either *Megalania* or crocs: modern kangaroos produce 48 teeth in a lifetime. Fossil mammalian teeth are even more common than fossil croc teeth, and the mammals—kangaroos and diprotodontids—did not live in or (presumably) near the watercourses, but of course they would have come to the water periodically to drink. Although there may have been some bias toward collecting mammalian rather than reptilian teeth, the experience of recent collectors suggests that this had little effect: fossil marsupial teeth really *are* more common. So even though there may not have been four hundred crocs to each *Megalania*—in fact, there almost certainly were not—*Megalania* probably was much rarer than the crocodilians, and this also is consistent with our estimate of their large individual range.

Finally, Auffenberg (1981) found that many individual oras were transients, often not fully mature, that passed through the ranges of resident oras over a period of a few weeks to a few months. Furthermore, the ranges themselves seemed to Auffenberg to be "vague" and were in a state of continuous flux (p. 175). Because such movement patterns (as well as the sizes of ranges) seem to be primarily determined by requirements for food, the presumed similarity in these requirements of both oras and *Megalania*, as well as their large size, leads us to expect that *Megalania* did likewise.

A Rosenberg's goanna (*V. rosenbergi*) weighing 1 kg eats about five times its body weight of prey per year (King and Green 1993, 84). Auffenberg found that a large ora (apparently about 45–50 kg) ate about 155–170 kg per year, about 3.5 times its own weight. Presumably, the relative amount of food needed decreases with increasing body mass, but obviously this can't go on to the point where the animal is so large that it does not need food at all. We don't know how active the Rosenberg's goanna was; it may have been relatively inactive. But the estimate for the oras was made for animals in the wild, so we should be able to use it to estimate the food required by *Megalania*. If these data are applicable to *Megalania*, and assuming that they can be plotted on a logarithmic scale, a 200 kg individual would require about 600 kg of food per year, and a 320 kg lizard about 900 kg per year (Fig. 5.17). For comparison, in the 1970s the Taronga Park Zoo (Sydney) reported that they fed their Tasmanian "devils," which weigh up to 8 kg, about 260 kg of meat per year (Collins 1973). This is 32.5 times their own weight. Schaller (1972) calculated that for African lions, the male requires 2,555 kg of meat per year, and the lioness, 1,825 kg. Given weights of 145 kg and 113 kg for the lion and lioness respectively, this is about 18 and 16 times their body weight per year. The estimates for *Megalania* given above range from about 2.8 to 3 times their body weight per year. These values reflect the decrease in metabolic rate with increasing mass discussed in the following chapter.

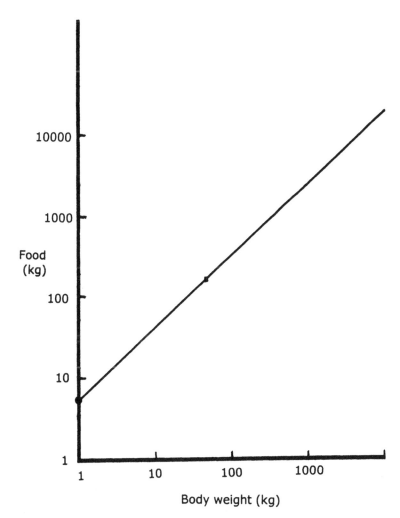

Fig. 5.17. Body weight vs. food weight for two varanids (Varanus rosenbergi and V. komodoensis). A, the datum for Rosenberg's Goanna; B, that for the ora, given as a box (very small at this scale) as a range of weights for the ora and its food were given. (Based on data from King and Green 1993, and Auffenberg 1981.)

In 1984 Peter Murray, a paleontologist with the museum in Alice Springs, Northern Territory, estimated the weights for Australian Pleistocene marsupials (Table 15). Given the digestive efficiency of monitors and their predilection to consume almost every part of the prey animal, the food requirements of a 200-kg monitor could be met by consuming one diprotodont, either *Diprotodon* or *Zygomaturus*, per year: those of a 320-kg individual by eating two diprotodonts. Presumably, the prey could be either hunted or scavenged. Given the large size of *Megalania*, these seem to be surprisingly modest needs, which contrast with the expense of being a mammal. If the lizards fancied kangaroo, Murray estimated the weights of three species as 50 kg or more, which would mean about ten to twelve 'roos per year, or one kill (or good scavenge) per month for our 200-kg lizard. Giant wombats probably fell in this size range as well. If the 320-kg animal favored large kangaroos such as *Procoptodon*, she could survive on even less than ten per year, possibly as few as four or five. These figures do not tax either our imaginations or the ability of the ecosystem to support *Megalania*.

Table 15
Estimated weights of extinct marsupials,
from Murray (1984)

| Animal | Weight (kg.) | Weight (lbs.) |
|---|---|---|
| Giant wombat (*Phascolonus gigas*) | 100–150 | 220–331 |
| Diprotodontid (*Zygomaturus trilobus*) | 500–1,000 | 1,102–2,204 |
| Diprotodontid (*Diprotodon optatum*) | 850–1,500 | 1,873–3,300 |
| Extinct wallaby (*Protemnodon anak*) | 40 | 88 |
| Extinct kangaroo (*Macropus ferragus*) | 150 | 331 |
| Extinct macropod (*Sthenurus occidentalis*) | 50 | 110 |
| Giant kangaroo (*Procoptodon goliah*) | 200–300 | 441–661 |

We have been considering adult *Megalania* and adult oras, but the young ones also had to eat. As mentioned above, oras select different prey depending on how old (and hence how large) they are. The young favor insects and bird eggs. But large ones can also select small prey, especially when times are tough. It would not be surprising if *Megalania* did, too.

However, if *Megalania* was so "cheap to run," why was it so rare? We can't easily (or confidently) compare our estimates of range size for *Megalania* with the estimates of how much an individual *Megalania* needed to eat, simply because we have no good idea of how abundant the prey animals were. Actually, we do know that many of the potential prey could not have been very rare because their fossils are not all that rare, but we don't know if there might have been one per square kilometer or a hundred. Even so, the estimated ranges and food requirements suggest that there should have been more individuals of *Megalania*.

One possibility is that these estimates are wrong. In fact, they are almost certainly wrong, but the possibility is that they are *substantially* wrong. They are, after all, estimates based on rather a lot of assumptions and rather a small amount of data. Furthermore, it just seems unlikely that any individual *Megalania* would always have her kill (or carrion) to herself. On Komodo, ora kills seem to attract other individual oras, and kills of other predators on other continents (Africa for example) not only attract other predators but also scavengers. Thus, our estimates of how much a *Megalania* needed to eat almost certainly underestimate their requirements.

The other possibility is that *Megalania* was not as rare as I have made out. In Estes's 1983 study of fossil lizards, he listed only twenty-eight specimens (from outside Australia). And this may be a slight overestimate, since it is a count of museum specimens, and museums may register as separate specimens bones that actually derive from a single individual (when this cannot be proven). Certainly many more monitors actually lived than that, so only a very small proportion become fossilized. It is reasonable to suggest that monitor remains do not fossilize well and that deriving the numbers from fossils substantially underestimates the number of individuals that actually did live. Why they may not fossilize well is not clear, although it has been reported that oras tend to scavenge carcasses of their fellows, and this may leave too little to stand a good chance of becoming fossilized. It also seems plausible to suggest that monitors, in general, may tend to die in areas away from where sediments are being deposited—such as rivers, estuaries, or caves. I think the suggestion that monitors do not fossilize well is almost certainly true, but this still does not mean that the estimates given here are not accurate or that *Megalania* was an abundant animal in its time.

## Social Life

When Ian Sobbe found the first known frontals of *Megalania*, the surprise was not that they were unexpectedly thick—for we knew that this might be a matter of strength—but that they bore a stout medial crest about one cm high (Fig. 3.3). At that time I did not know that any monitors had crested skulls; however, one ancient relative, *Telmasaurus*, had a low median crest on its skull roof (Fig. 4.11). Another, less-close relative from Cretaceous Mongolia, *Paravaranus*, had a low nasal crest at the base of its snout—and perhaps along the whole snout. But unfortunately the snout is broken on the only known skull. On others, the roof of the skull is reasonably flat. The crest is important for understanding the biology of these animals because it is not a feature useful in catching prey. Instead, like other such features, it was presumably useful in "catching" a mate.

Although many biologists have doubted its efficacy and even its existence, Darwin's notion of sexual selection is a significant factor in animal life and hence an important mechanism of evolution. Sexual selection shapes features that help organisms to success in mating. There are two major kinds of features that assist males in successfully persuading females to be mates: those that serve as weapons in competition with other males, and those that are attractive to females. Unfortunately for the clarity of the concepts, each is often the other.

Weapons used to fight with or to intimidate other males are often attractive to females. The antlers of stags are an example here, or the jaws of stag beetles, but not the tails of peacocks. (The tails may indeed intimidate rival males, but they are not weapons.) Females apparently notice those features of males that signal the strength or vitality (health) of the male and select those males as mates that are indicated as being the strongest or healthiest. Much fascinating work has been done on this subject, and good places to learn about this are the books of Andersson (1994) and the Zahavis (1997).

Interpreting the cranial crest of *Megalania* as a secondary sexual feature—primary sexual features are those functioning directly in sexual reproduction—does not automatically clarify whether it was a weapon or a signal. It could have been either—or both. We could test the idea that it was an indicator of vitality if there were a sufficient number of skeletons, since you can sometimes determine the state of health of the animal from the structure of the bone. If well-developed crests correlated with healthy—or at least healthier—skeletons, and poorly developed crests with less healthy ones, this would be good supporting evidence. With the small amount of fossil material there is of *Megalania,* we have little serious chance of doing this. However, a third frontal has recently come to light near Townsville on the central Queensland coast. This frontal reportedly has a relatively lower crest than those from the Darling Downs, so crest size seems to vary as expected from this interpretation. Furthermore, if the crests were weapons, we might expect to find some injuries, probably to the ribs, that resulted from head butting.

Amazingly—in view of the general lack of fossils of *Megalania*—we do find such injuries. The specimen from Alderbaran Creek has one rib that was broken and then healed. The break is near the lower end of the rib, and the broken piece was not displaced. It healed in the proper position but with the development of a thickened "knot" of bone at the break, known in medical circles as a callus. Usually, with time, calluses are resorbed, so this may indicate that the injury occurred not long before the death of the animal: long enough for healing to take place, but not long enough for the callus to be lost. But again, we must be cautious about the implications of this injury. Maybe it resulted from combat between male *Megalania,* or maybe it was the result of a swift kick from an animal that disputed becoming the *Megalania*'s dinner. It does not prove the use of the crest as a weapon, but at least it is consistent with that notion. However, this is not the only such specimen: there is also a broken and healed rib in the remains sent to the Australian Museum by Lau. If there are really only some twenty individuals known, as we suggested in chapter 4, this indicates that broken ribs, although uncommon, were not very rare.

Living varanids don't engage in head blows during fighting, although other lizards, such as the marine iguanas of the Galapagos, have head-to-head shoving matches. *Megalania* may have done either head butting or head-to-head shoving (or both)—or, as suggested above, may have butted the flanks of an opponent. As a weapon in either blows or shoving, the crest would act to concentrate the force of a blow with the head and thus would inflict more damage than a blow from a skull lacking such a crest. A horn would concentrate the force even more, which is presumably why horns evolved in many different kinds of animals. The crest would not have been particularly effective as a weapon because it does not stand high above the level of the top of the skull. Thus, if the head met its opponent even slightly less than directly head-on, some other part of the skull roof would also contact the opponent and spread the force of impact (or shove), reducing its effect. But this is no reason to believe that the crest was not used as a weapon, because even some small effect in deterring one's rivals is, on average, better than none. Furthermore, the height of the crest may have been increased by a scaly or horny sheath.

As a display structure, however, the crest is well placed. The attention of animals (and people) is often directed to the heads of other animals (or faces of people). And the top of the head of what is, after all, a fairly low-standing quadruped is probably the most prominent place for a display structure.

We cannot be sure that the crest was used either as a weapon or as a signal. But unless we are very far wrong about the processes of evolution, we can be sure that it did have a function. After all, if it did not, those individuals of *Megalania* that did not waste metabolic energy in growing a crest, even a small one, would have an advantage when food was scarce—which must have happened from time to time. Gradually, they would have come to dominate the population and the crest would be lost.

When the frontals and parietals of *Megalania* were found, as already mentioned, it was quite obvious that they were thick, massive bones. The corresponding bones in living varanids are not remotely as thick, even taking into account that their skulls are much smaller. Was this a clue to some unexpected behavior or adaptation of *Megalania*? The discovery of a cranial crest suggested that *Megalania* may have indulged in head butting—so perhaps this thickening was the result, developed to protect the head? This seems like a good explanation, but it is not the only, or even the simplest one.

We have known at least since Galileo that if the dimensions of a bone are doubled, the larger version is weaker for its size than the original. The strength of a bone depends on the area of its cross section. Doubling the length and other linear dimensions (such as width) increases its cross section by four times (in mathematical terms, the square of the increase in linear dimensions). So far, so good—in fact, too good. Bones support the weight of the beast, and weight depends on volume. So the length increases by two times, the strength by four times, and the weight by eight times (the cube of the increase in the linear dimensions). To keep pace with the increased weight, the bone has to increase in cross-section by eight times, not four: its increase in width has to be at least twice its increase in length.

We have tacitly assumed that we are considering a limb bone, but this argument also holds for the bones in the skull. Physicists and anatomists have worked out methods of calculating how the different dimensions of a bone must change in order for it to maintain its strength with respect to size as it increases in size. McMahon's method of elastic scaling (McMahon 1973; McMahon and Bonner 1983) indicates that the square of the diameter of a column, such as a limb bone, will equal the cube of its length. The frontal of *Megalania* is not a column but (roughly, at least) a plate. For a plate we can equate the depth times the breadth to the cube of the length (or in mathematical terms, $D \times B = L^3$). When the frontal of *Megalania* is analyzed in this way (Molnar 1990), it turns out that the obvious thickness is simply there to maintain its strength. If a frontal (I used QM J14498) of the smaller crocodile monitor (*Varanus salvadorii*) is mathematically enlarged to equal in length that of *Megalania*, elastic scaling predicts that to retain the same proportional strength, its thickness should become 20 mm. The better preserved of the frontals (QM F16783) of *Megalania* is 19.8 mm thick. So the thick bones of the roof of the skull were there to protect the head all right, but to protect it from its own increased weight, the same increase in weight that dictated the loss of the flexible joints in the skull.

But there is one other feature to consider. Their thickness and their possession of a crest are not the only unusual features of these frontals. The upper surface of the bone is covered with low but distinct ridges rather as if thin strands of spaghetti had been laid on top of the frontal and become part of the bone. These are not found in other monitors, either living or fossil. Whatever the origin of these ridges, they would have acted to hold the skin in place on the bone's surface and prevent it from slipping or tearing where the dermis attached to the bone. If the skin of the top of the head never came into contact with anything, there would be no advantage to holding the skin more firmly than usual for other monitors, and the effort of growing the ridges would be wasted effort. This suggests that the skin was subject to sliding or slipping, which in turn suggests that head butting or shoving did take place.

This all indicates that among *Megalania*, as among other animals, males competed for the attentions of females. This should come as no surprise. But why head butting or shoving? Large modern varanids don't do this; instead, they engage in a wrestling matches while standing upright on their hind legs if they come to actual combat (they may also bite). Auffenberg's team never observed this behavior in oras, but a Japanese television crew videotaped it. Naoki Suzuki, at the Medical Engineering Laboratory of Jikei University (Tokyo), and his colleagues (1991) analyzed the images. They found that the lizards, when two were braced against one another, could remain upright for "more than a couple of minutes." They also observed that oras could stand upright on their own for up to about 20 seconds when retrieving bait hung from a branch. Unfortunately, they did not say how long these lizards were.

When an ora stands upright, the femora are inclined at about 45° to the horizontal. This is not a particularly good stance for efficient structural support. Possibly the ancestors of *Megalania* also used this mode of combat but outgrew it and became too large to successfully stand upright. Then biting or some other activity of the head was adopted in its place, and this provided the opportunity for the evolution of the crest. Since small varanids wrestle using all four legs, the bipedal stance in the large monitors may also be a result of their increased size, lost in *Megalania* with its even greater increase in size.

It has generally been thought that monitors do not show sexual dimorphism in its strict sense of a difference in form between the sexes. However, Graham Thompson of Edith Cowan University and Philip Withers of the University of Western Australia, both in Western Australia, found that Australian monitors generally do show some sexual dimorphism (Thompson and Withers 1997). Males have larger heads, longer limbs, and shorter trunks than females, although this is not immediately apparent to the eye and required many measurements and much statistical analysis to discern. It is not as far-fetched as it may seem to propose that male *Megalania* had a crest, but females did not. Even so, in some ways its biology would have been different from that of any living varanid.

Finally, we should recognize that this section on social life is more speculative than the others (except, perhaps, for that on the den). This is not because of the considerations of sexual selection and the concomitant fighting for, and choosing of, mates but because the crest is known from the only three frontals that have been found. Thus, we do not *know* that the crest was dimorphic; we just assumed it. To know that requires the discovery of one more well-preserved frontal, lacking the crest or having a more subdued one. It is true that the crest of one of the frontals is substantially less prominent than that of the other. However, the frontal with the more subdued crest is also the more badly worn, so this may just reflect its having been abraded before fossilization. If the crest was not a dimorphic feature, then there is no plausible explanation for either its existence or its evolution.

## Reproductive Strategy

Sexual dimorphism in varanids is not always obvious, but often there is a difference in size that is. It has long been difficult to distinguish male from female oras, although Auffenberg reported a difference in scale patterns near the vent. No *Megalania* scales (much less scale patterns) have been found, so this is not helpful—if it even applied to *Megalania*. The remains of *Megalania* are too sparse to determine if some individuals had relatively larger heads or longer limbs—although if *Megalania* is closely related to the other Australian monitors, this probably would have been the case. The largest *Megalania* were probably males, and the smaller may have been females (or young males). On average, female oras were about 78 percent as long as males, but Auffenberg also reported that the largest female ora was 94 percent as long as the largest male he found.

Australian monitors usually breed once a year, probably because breeding requires energy, and the Australian climate is either too cool or too dry for the energy to be available in the winter. Tropical lizards, where the climate is more benign, may breed all year, but oras usually do not. During much of the Pleistocene, the Australian climate was cooler or drier than now, so we would expect the same of *Megalania*. However, animals do tend to take advantage of "good times," and it

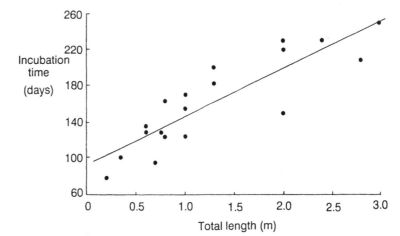

Fig. 5.18. Incubation time (in days) vs. total length (in meters) for varanids. (From King and Green 1993.)

would not be surprising if *Megalania* bred more often during the periods of less rigorous climate. Oras form pair bonds, so the ratio of breeding males to females is about 1:1. Presumably, this was also the case for *Megalania*.

King and Green (1993) present graphs of incubation period as related to total length, and of clutch size (number of eggs laid at one time) as related to total length for varanids (Figs. 5.18, 5.19). These can be used to estimate incubation period and clutch size for *Megalania*. For a 3.5-meter female, the clutch size is 25 eggs; and for one 4.5 meters in length, it would be 36. On average only two (or fewer) young of these would be expected to have survived, otherwise fossils of *Megalania* should be much more common. The incubation period for a 3.5-meter female would be about 280 days, and for one of 4.5 meters, approximately 330 days. Checking this with the graphs of Thompson and Pianka (2001) gives a reasonable match for the estimated incubation period of about 235 and about 250 days respectively for 3.5-meter and 4.5 meter females. Estimated clutch sizes are somewhat larger, 63 and 80 eggs respectively. None of these seems to be obviously implausible, but we must emphasize that *Megalania* probably reproduced relatively slowly, as is generally true of large animals. What we do not know is whether *Megalania* followed this trend, or perhaps laid fewer but relatively larger eggs. Larger eggs could have produced more precocious hatchlings and hence (presumably) increased the chances of their survival. The large size of *Megalania* implies that large eggs could have been produced and laid, but the only way to tell is to inspect the shells. Unfortunately, all eggshells so far found in the Pleistocene beds of Australia derive from birds.

More important than the time taken to reproduce is the energy required. Rosenberg's goanna, the only one for which I have been able to find data, requires an increase of 32 percent over the average daily intake of food. If this were also true for *Megalania*, a female of 200 kilos would require (approximately) an additional 200 kilos of prey to

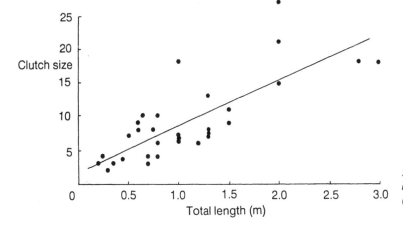

Fig. 5.19. Clutch size vs. total length (in meters) for varanids. (From King and Green 1993.)

reproduce (lay a clutch of eggs), and one of 320 kilos would require about 300 kg more food to do so. Consulting Murray's estimates (1984), this could be provided by one kill of a large kangaroo or small diprotodont. In view of the rather low number of kills estimated to have been required to support a *Megalania,* this does not seem to present a problem—unless prey were really scarce. A very large female of about 2,000 kg, however, would require a very considerable 2,500 more kilos, the equivalent of two *Diprotodon.* However, if the very large specimens were males—as suggested earlier—this is irrelevant.

## Science and Fiction

At the end of his 1976 history of science fiction, David Kyle wrote, "science fiction is alive and well and living in the scientific community" (p. 169). In a serious sense, what I have just written is science fiction. We cannot observe live *Megalania,* so we must reconstruct in our minds how these creatures lived. We do not have complete information about them—that is hard to come by even for living creatures—so we must conjecture as much as we can of what we don't know. How seriously do we take these scenarios?

In short, as seriously as we can take the evidence. These are not sheer guesses but are based on five lines of reasoning: (1) extrapolation from modern data; (2) consistency with such fossil evidence as we have; (3) consistency with what is known of modern relatives; (4) consistency with modern analogs; and (5) consistency with theory. The second of these is the best (or "most robust," in the jargon) because it is direct evidence. Fossils were formed, almost literally, by the forces that shaped the lives of the creatures from which they derive. This is true both in the indirect sense that natural selection by environmental factors has shaped the form of the creature encoded in its genes, and in the direct sense that bones reflect environmental influences in how they grow and whether or not they are injured.

The third line is probably the least reliable because good reasons why closely related animals should live in similar ways have never been found. Birds and crocodilians are more closely related to each other than either is to any other living group, but no paleontologist would attempt to reconstruct the life of an extinct croc using a modern bird as a guide. This line is used here, however, because varanids are a special case. The conservatism of body form in monitors means they are both the closest living relatives and the best living analogs. We have especially used the ora in this fashion.

The fifth line, consistency with theory, may sound strange to those who feel that evolutionary biology cannot be a predictive science. As we saw in the case of Hecht (1975), some limited predictions can be made and, with the luck of finding appropriate fossils, can be tested. And really, prediction is always limited by the degree of our understanding—which is why the unified field theory that Einstein searched for (now called a "theory of everything"), relating gravity to the forces of electromagnetism and nuclear interactions, still eludes us half a century later. However, in paleontology our understanding may also be limited by the lack of fossils. On the positive side, careful experimental and observational research in evolutionary theory has verified that many evolutionary processes, such as competition, that were once merely *thought* to be important, really *are* important. This gives us the confidence, for example, to consider the possibility of sexual selection in *Megalania*. But the theory used is not just that of evolution; what we know of physiological ecology is also helpful: that large animals need more food than smaller, for example, and hence require a larger range.

Still, some humility is in order. We need to remember that we may be wrong, that in its evolution *Megalania* may have "thought" of something that we haven't. So this is science fiction in that we have no direct evidence for much of it—the fiction; but it is based on what we know of fossil and living organisms and their evolution—the science.

# 6. Why *Megalania*?

## The Evolution of Varanids

Giant lizards are not common today, nor have they ever been common since lizards first appeared. By "giant" I refer to lizards that reach a length of three or more meters—in other words, comparable in size to moderately large mammals, lizards that are giant relative to us, not animals like the giant chameleon that are giants only relative to other chameleons. Naïvely, we might expect that giant lizards would be found during the times of other giant reptiles such as the dinosaurs. However, except for marine lizards (the mosasaurs) this just is not the case. On land, giant lizards (and giant tortoises) only occur after the dinosaurs became extinct.

We tend to think, because mammals succeeded the dinosaurs as the large, prominent land animals, that somehow the mammals were better equipped to survive than reptiles. After all, until recently, when it was realized that birds were descendants of dinosaurs, it was thought that the dinosaurs had vanished completely. And mammals do have advanced anatomical, physiological, and behavioral characteristics, so they should be formidable competitors to any other land-dwelling tetrapods. If we see lizards at all, it is usually briefly, as they disappear under bushes or into cracks in rocks—hardly creatures as formidable as mammals. But apparently those perceptions (or preconceptions) are wrong; otherwise giant lizards would not have appeared during the Cenozoic and still survive today.

Thus, the question of how *Megalania* came to be, of what evolutionary influences resulted in this creature, is an important one, both because such creatures must have played an important role in the habitats in which they lived and because the very existence of such creatures was unexpected. This question we shall consider here.

Not only is the lizard-like form itself ancient, but the specific form of the monitor lizards is also old. As we have seen, lizards that probably looked like modern monitors lived as long as 80 million years ago, in the times of the later dinosaurs. In gross features, the anatomy of these lizards is still with us. Few other tetrapods have changed so little in external appearance—although frogs and salamanders and some crocodylomophs are at least as old and as little changed. Monitors—although advanced in some of their physiological features—are quite conservative in their external forms.

Frogs have a broad range of biochemistries, some of which must be the product of substantial evolutionary change. But we would easily recognize frogs from 190 million years ago as frogs, perhaps even mistake them for modern ones, could we but travel back in time to see them alive. Presumably it is this repertoire of biochemical abilities that accounts—in part—for the long survival of frogs with little anatomical change. Of course, we know almost nothing of the physiology of ancient monitors, but we can take the example of frogs as an inspiration to speculate how the physiological features of modern monitors may have permitted this group to survive for so long with little change in form.

The monitor body form must, in a sense, be multifunctional. Many monitors live on the surface of the ground, but others are aquatic, and yet others—generally small ones—live in trees. Those that live on land can inhabit a variety of habitats, including forests, mangrove swamps, plains, and deserts. Most are carnivores (and insectivores), but the Philippine Gray's monitor, *Varanus olivaceus,* enjoys a good meal of fruit. This suggests that the form can readily accommodate to several different modes of life without obvious external change. Gray's monitor, however, exhibits some notable internal changes: the large intestine is enlarged, with a cecum; the skull and jaws are unusually robust for a monitor; and the posterior teeth are blunt in the adult (Auffenberg 1988). The recent discovery of a second herbivorous monitor, *Varanus mabitang* (also in the Philippines) emphasizes the potential versatility of these lizards (Gaulke and Curio 2001; Struck et al. 2000).

In frogs, the body forms reveal little variety, but their biochemistries vary widely; in monitors, the body forms also show little variation, and size (mass) corresponds to the frogs' varying biochemistry. The largest living monitors weigh 100,000 times as much as the smallest (Pianka 1995). If we include *Megalania,* the range is even greater. Mammals and dinosaurs (including birds) show similar ranges in weight, but both groups consist of hundreds or thousands of genera, whereas living monitors form a single genus. No other genus of land-dwelling animals approaches this range in size. The anatomy and physiology of monitors seems to be suited to a wide range of sizes (masses), as well as to having an impressive "shelf life." Eric Pianka (1995), of the University of Texas at Austin, investigated the evolution of body size in varanids. Unfortunately, the lack of information about which of the modern forms are the most primitive prevented him from reaching any interesting conclusions. Many of the oldest monitors and related lizards are quite big, but this may just demonstrate that large animals are more easily preserved in the fossil record than small ones (which is known to be the case), rather than that the early monitors were all large animals.

The work of Fuller and colleagues, however, indicates that the African varanids are the most primitive (1998, Fig. 4). Combining this with the 1995 study of Pianka, we can sketch out how size probably evolved in monitors, although bearing in mind that only twenty-one of the forty-five species of monitors were studied by Fuller and her team. When more monitors are studied, these conclusions may require re-evaluation. It appears that early monitors were rather large, around a meter (3 feet 3 inches) long overall. The Australian pygmy monitors appear to be a natural group of small varanids, but only if Gould's monitor and the perentie are included as members of this group that became large—"giant pygmy monitors," so to speak. The emerald monitor (*Varanus prasinus*) of New Guinea, on the other hand, apparently derived from a large ancestral lineage and became small. The oras seem to be giant descendants from a lineage of large monitors. This study also supports the old contention that oras have Australian relationships, since the ora forms a natural group with the lace monitor of Australia and the crocodile monitor of New Guinea.

Several of the distinguishing features, four out of eight, of modern monitors (the Varaninae) involve the structure of the vertebral column. At least one of these features almost certainly relates to a lengthening of the neck. To understand, or at least speculate about, the significance of this, recall the discussion in the preceding chapter on breathing in monitors. The gular pumping of monitors involves the "swallowing" of air to force it into the lungs, rather than sucking air into the lungs by expanding the rib cage, the only method used by other lizards. Increasing the length of the cervical (neck) vertebrae presumably increases the length of the neck and so also increases the volume in the trachea available for air for the "gular pump." Thus, a longer neck suggests the ability to force more air into their lungs than their ancestors, and hence the possibility of greater activity levels and stamina.

However, features of animals are rarely used for a single purpose: dogs use their forepaws to walk or run, to dig, and to hold tasty bones to gnaw. Walking is presumably the dominant function here, but the others must also have some effect on the life of the creature that may be reflected in their evolution. Similarly, the longer necks in monitors must have conferred some advantage. Losos and Greene (1988) were impressed by the ability of monitors to capture fast-moving prey animals and attributed this to the mobility of their heads, which in turn is due—in part—to their longer necks.

## The Evolution of Giant Varanids

Based on our reconstructed biology of *Megalania*, we can try to find out, deduce, and reconstruct what we can of the influences that resulted in the evolution of this creature, that is, the factors that governed the evolution of giant varanids. Given the efficiency of digestion in varanids, or at least their ability to devour most of a carcass and so make it available for digestion, we can suggest that the ancestors of *Megalania* were already in a good competitive position vis-à-vis the marsupial meat eaters. When food was hard to find, the monitors could more effectively utilize what they did kill (or find). Their lower metabolic rate added to their advantage: remember that Rosenberg's goanna needs five times its body weight in food per year, while the Tasmanian "devil" needs just over thirty-two times its weight (Fig. 6.1). A large monitor could make do with less food—first, because the food was (probably) used more efficiently; and second, because less was needed to adequately fuel their metabolism. This may help account for why monitors were around at all, but not for why some became big.

In oras, large body size permits the animals to maintain their body temperature more nearly constant, in spite of changes in the temperature of their environment, than their smaller relatives can (Auffenberg 1981). This is not to say that their body temperature is constant, just that they do "better" than smaller monitors. Thus, they can be active more often and for longer periods than smaller monitors. Work by James Spotila, from Drexel University in Philadelphia, and his colleagues Michael O'Connor (also at Drexel), Peter Dodson (University of Pennsylvania), and Frank Paladino (Purdue University) in 1991 may be relevant in examining this aspect of the biology of *Megalania*. They modeled the thermal characteristics of dinosaurs. The estimated weight of one of the animals they simulated, *Tenontosaurus*, was 624 kg, about the same as Hecht's estimate for *Megalania*, and close enough to our estimate to be useful in this context. The general body form was not greatly different—*Tenontosaurus* had relatively longer hindlimbs and tail—but this form was clearly more similar to that of *Megalania* than those of, say, a tortoise or a large mammal. Spotila and his team calculated that with a "typically reptilian metabolic rate," *Tenontosaurus* would have been able to maintain a body temperature 1–2°C above the environmental temperature. If it raised its metabolic rate tenfold during activity, the body temperature could rise to about 10–20°C above that of the environment. Given that the metabolic rate decreases with increasing body size, that of *Megalania* may plausibly have been close to the "typically reptilian" rate the team used. Actually, the rate does not decrease—as we shall explain shortly—it just increases less rapidly than we might expect. And given the observation of Bartholomew and Tucker (1964) that the rate in active monitors could approach or surpass the basal rates of mammals, the active metabolic rate used in the model is also plausible.

However, we must realize that this piece of "science fiction" is a little more fiction than those of the previous chapter. There we looked at data for modern monitors and used it to guide our visualization of *Megalania*. Here we really have no data; the study of Spotila and his colleagues was—in a sense—pure mathematics, a simulation on a computer. This simulation may be quite accurate, reasonably accurate, a bit off, or wildly wrong. Although I think that the last alternative is probably incorrect, we cannot be 100 percent certain of the results. Nonetheless, it's the best we've got, so we will assume that it is close to accurate.

If *Megalania* could generate a body temperature of a few to, say, 15° above that of its surroundings, it could presumably maintain it in the "mild" (i.e., not freezing) climate of Australia by virtue of its large body size. The study by Spotila and his colleagues indicated that

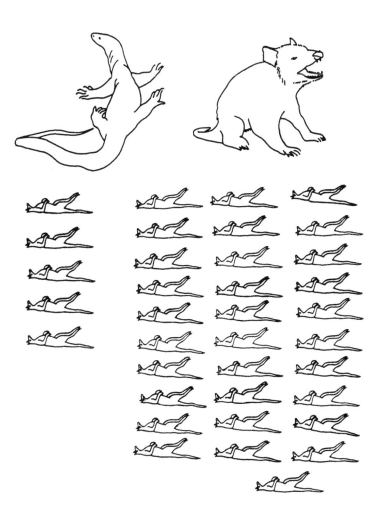

Fig. 6.1. The amount of food eaten by an ectothermic predator (Rosenberg's Goanna) compared to that required by an endotherm (here the Tasmanian "devil"). Each kangaroo carcass represents the body weight of its predator, thus those under the goanna are not equivalent to those under the "devil." Still the point is clear: endotherms require more "fuel" than ectotherms.

*Tenontosaurus* would have taken about two days to respond to changes in environmental temperature. So if their analysis is correct, reptiles of this size would have been able to weather daily cycles of temperature with little change in their internal temperature. In the cooler climate of (parts of) the Pleistocene, this may have given *Megalania* a substantial advantage over other reptilian predators (but presumably not over marsupials) as long as the temperature fell and rose for short periods, say nightly, but did not stay cold for a week or so. Large individuals that don't chill off at night can get up and about and hunting earlier in the morning and can hunt later in the evening than can smaller ones. Increasing the time available for hunting should also increase their chances of success. Modern Australian monitors generally are not active during the winter, except in the far north, but *Megalania* was presumably big enough to be active through at least some of that season as long as, like in modern Australia, the days were moderately warm. Once a *Megalania* heated up, she would remain warm until there was a cold spell lasting more than two days. And heating would presumably not be a problem during summer. If or when she did cool off, her metabolic rate would fall, and she would need less food until she warmed up again (like the smaller monitors).

Unfortunately, we don't know enough about the Australian climate during the Pleistocene, for either the glacial or the interglacial periods, to have any serious idea of how much colder (or warmer) it was, or of the range of daily and seasonal temperatures. The conclusion that *Megalania* could have been active for at least some parts of the winter, and more active than their smaller relatives, thus seems reasonable. As for the summer, we might even suspect that heating up was less of a problem than cooling off. Although "summers" in northern Australian are often cloudy—this is the wet season—clouds would not have appeared every day, and clear spells are known. Wet season or not, sunlight is sunlight, and the wet season is summer, so the sunlight is more intense. At times during the wet season there would have been a lot of water on the ground that would have disappeared during the dry season, and this could have been used for cooling, just as oras cool down by lying in pools. These considerations assume that the climatic conditions during interglacials, at least, were similar to those now in that region.

But we know that *Megalania* did not evolve its large size because large individuals had an advantage during cool or cold weather and passed that on to their offspring. It was not selected for large size by the deteriorating temperatures of the Pleistocene climate. We know this because there were already large individuals in Early Pliocene times around 4 million years ago. Although there was a decrease in temperature early in the Pliocene, it was not so much as later during the Pleistocene. *Megalania* was already large, so it just happened to be "ready" for the cooling climate.

Hunting is a possible factor to consider in the evolution of large size in the ancestors of *Megalania*. The larger the hunter, the larger the prey it can subdue. On the other hand, the larger the predator, the more prey it must subdue. At first we might think that being large enables a predator to catch and eat more prey. A moment's reflection, however, suggests that since the predator is larger, it must eat more to stay alive than it would if it were smaller. Is there any profit in this? In other words, by becoming larger can one catch sufficiently many more prey animals that this increased "catch" offsets the increased requirements for food, the increased hunger? The answer may be "yes." Physiologists have found that for closely related animals, the greater the beast's mass, the lower its metabolic rate (when resting at least). Actually, it is metabolic rate per kilogram (or any unit of mass) that decreases. Thus, one elephant is cheaper to feed than one "elephant-weight" of mice. This seems to hold not only for endothermic creatures (those that generate their body heat) but also for ectothermic ones (those that do not generate their body heat, but absorb heat from their environment)—at any rate, Bartholomew and Tucker (1964) showed that it is true for living monitors. So a *Megalania* should be cheaper to "run" than a "*Megalania*-weight" of, say, lace monitors.

In trying to understand how and why the really big meat-eating dinosaurs like *Tyrannosaurus* became so large, Jim Farlow prepared graphs for how big a predator can be supported by prey animals of some given size (several given sizes, actually) (1993, Fig. 3). One also has to consider how many individual prey animals there are for each predator. Given ratios of 10–20 prey per *Megalania*, large kangaroos weighing 200–300 kg (as estimated by Murray) can support predator weights of 200–700 kg for *Megalania*. If we have 50 prey animals per *Megalania*, then *Megalania* could reach 1,500–2,000 kg. These numbers of prey do not seem extraordinary—especially in view of the rarity of fossils of *Megalania*—so there seems no obvious bar to the evolution of giant monitors from the availability of potential food in Australia during the Pleistocene. Auffenberg found that ora populations are not limited by the availability of food.

Endothermic predators require about ten times as much food as ectothermic predators of the same weight. But no one has considered monitors—living or fossil—to be endotherms. And there is some work on estimating body temperature for a Cretaceous North American monitor, from geochemical analyses, that indicates that the animal was ectothermic (Barrick, Showers, and Fischer 1996).

Another feature discussed by Farlow is the "turnover rate" of the prey animals. This is the rate at which new individuals grow up and enter their population. Since *Megalania*, like oras, probably did not restrict its hunting to adults, we might simply consider the reproductive rate of the prey. Here we find that marsupials are more "flexible" than placental mammals: marsupials can adjust their reproductive rates to match environmental (especially climatic) conditions. Not only can they control how much effort (or "investment") they put into developing and raising young, but they can suspend development when environmental conditions are poor (Dawson 1983). This enables them to control their turnover rate, decreasing it to zero when times are hard, or—for modern kangaroos—increasing it to three per year when conditions are good enough to support that rate. *Megalania* might well have been suited to this kind of prey. Being ectothermic, it could presumably fast through the periods when conditions were poor and profit from the rapidly increasing marsupial population when conditions became good. We don't know if this reproductive strategy was also used by diprotodonts, but presumably it was used by the fossil kangaroos.

So was the evolution of large size in *Megalania* driven by the greater hunting (or least feeding) success of larger individuals? In oras, large individuals do have an advantage in competition with their smaller fellows at carcasses, and it does not tax the imagination to think this of *Megalania* as well. But another factor may enter into an account for its large size. For that, we need to turn to Auffenberg's (1981) comments on the role of sexual selection in oras (similar observations have been made by Bennett 1998).

Auffenberg noted that the mating system of oras "seems designed to select the most powerful mates" (1981, 176). And in monitors this generally means the largest. This mating system is consistent with the evidence suggesting that sexual selection was important in *Megalania*. If varanids arrived in Australia during the Miocene, then given their digestive "virtues" discussed in the preceding chapter, they may have faced relatively little competition from the meat eaters among the Australian marsupials and other lepidosaurs. Sexual selection could have operated unhindered to produce a large body size, which was then helpful both in the deteriorating climate of the Pleistocene and for acquiring larger prey.

It must be added, however, that an absence of competitors seems to be an unnecessary factor for varanid success. John Phillips, from the Zoological Society of San Diego, found (1995) that the biomass of the white-throated savanna monitor (*V. albigularis*) in the Etosha National Park (Namibia) was as great as that of any of the mammalian predatory species (which include lions, leopards, and spotted hyenas). This shows that moderately large monitors can thrive even in regions also inhabited by mammalian predators of moderate to large size.

Against the influences of digestive efficiency and sexual selection stood the architecture of the limbs of these monitors. They have not achieved an erect stance as both mammals and dinosaurs did. Possibly this was also related to the lateral flexion of the spinal column in walking. But they did manage to free themselves from the constraint on breathing. They may have been on the way to an erect posture: videos of oras show that when moving rapidly, they pull their legs as close to underneath them as they can. We know from the history of dinosaurs (and some mammals, such as the indricotheres, giant Asian rhinos of the Oligocene) that an erect posture permits the evolution of very large animals. Since an increase in body size in the *Megalania* lineage was probably not inhibited by the availability of prey, perhaps the sprawling stance kept them from becoming any larger than they were.

It is not clear which—if any—mammalian carnivores competed with *Megalania*. However, marsupials would be able to produce young in less than the 330 days of a large *Megalania,* if only because they are smaller. Furthermore, with embryonic diapause—the ability to indefinitely suspend development of a fertilized embryo (also termed delayed implantation)—marsupials can hold their embryos through harsh environmental conditions and then quickly bring the offspring to independence with the appearance of good conditions. This gives them a head start in taking advantage of good conditions to raise more young more quickly than a placental mammal or a lizard, because they are not starting from mating, like most other animals, but already have a partially developed young. A *Megalania* would have to wait for almost a year—on this model—before her young appeared. Thus, marsupials are better suited to dealing with uncertain environmental conditions.

But whatever the result of these considerations, they shed little light on what actually happened, because the large carnivorous marsupials and *Megalania* both became extinct. Whether or not competition (or its absence) played a role in the evolution of *Megalania*—and we shall argue that it may well have had a role—it seems not to have done so in its extinction, at least not obviously.

Populations of *Megalania* probably were not limited by the availability of food. Murray's (1984) estimated weights for Pleistocene Australian marsupials indicate that some were substantially bigger than living marsupials. The largest modern kangaroos weigh up to 60 kg; extinct ones weighed up to 200–300 kg, and diprotodonts to maybe 1,500. Flannery (1994) argues that all marsupials weighing more than about 100 kg became extinct by the end of the Pleistocene. Given this, it becomes apparent that *Megalania* probably succumbed to the lack of prey when the large marsupials died out. This may well have been due to human interference: what would have happened had humans never arrived, we do not know.

## The Geography of Giant Lizards

Large monitors are concentrated in the lands of the southwestern Pacific region: Australia, New Guinea, Indonesia, and the Philippines. Based on the dimensions given in the book by Bennett (1998), all but two of the twenty monitors with a snout–vent length of 25 cm or more come from this region, as well as twenty-nine of the thirty-one with a total length of more than a meter. Moreover, all monitors with an overall length of more than 1.5 meters live here. (The different numbers of monitors for which the two measurements are available result from some people measuring only snout–vent length and others only total length.) Thus, it does not seem out of place that *Megalania,* the largest monitor, was also found here. So we may ask why: Is there something unusual about this region? What factors here caused or permitted the evolution of such lizards? We may ask—but answering may be another matter altogether.

A confounding factor is that it is not just most *large* monitors that occur in this region, but most of any kind of monitor: about two-thirds of all monitors live here (Cogger 1983). The smallest varanids, the pygmy or dwarf monitors, also occur here. However, most monitors do not regard large vertebrates as prey (Losos and Greene 1988). Since it is in one of these exceptions—*Megalania*—that we are interested, we will look specifically at big monitors that do feed on large mammals.

The significance of climate for evolution is increasingly recognized, so one relevant influence may be the climate. We generally picture Indonesia as a lush, tropical paradise of rain forest—at least primordially. Or, depending on one's attitudes, a dense, formidable tropical jungle, the "green hell" beloved of feature writers fifty-odd years ago, who were probably influenced by the experiences of soldiers and reporters in the Pacific theater during the war. But the country of Indonesia is large, and the Malay Archipelago larger still. Rain forests are indeed found in the islands of Kalimantan and Java, among others in the west. But the islands of the Nusa Tenggara, or Lesser Sundas, which include Komodo, are the driest in Indonesia. Here the rain is highly seasonal, falling from December to March. This is also the climatic pattern through much of Australia, especially in the north.

Australia has an unusual climate. The diversity of weather experienced at any one time in North America or Europe is absent here. The continent lies centrally in the Indo-Australian plate, so the high mountain ranges—which tend to form at the edges of plates—are in New Guinea and New Zealand. Australia itself has no substantial mountains to cool the passing air and precipitate rain or to break up weather

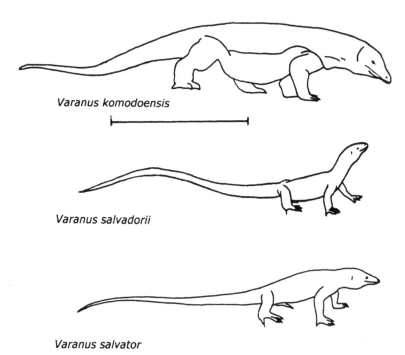

Varanus komodoensis

Varanus salvadorii

Varanus salvator

Fig. 6.2. The living giant monitors to scale. Top, the ora. Middle, the crocodile monitor, Varanus salvadorii. Bottom, the water monitor, Varanus salvator. Both the crocodile and the water monitor have proportionately smaller bodies and longer tails than the ora. Scale bar is 1 meter.

patterns formed over the vast flatnesses of the Indian or southwestern Pacific Oceans. Furthermore, Australia is longer from east to west than from north to south. Since climate, like average temperature, tends to change as one approaches or recedes from the equator, much of the continent lies in the same climatic zone. As a result, broad regions experience the same climate and the same weather. Rarely is the weather the same in London, Rome, and Moscow or in Tampa, San Francisco, and Chicago, but the weather in Alice Springs yesterday is often the weather in Brisbane tomorrow. The climate of the northern part of Australia tends to be dry, with a "summer" wet season. In short, the climate is basically similar in both the Komodo region of Indonesia (the Nusa Tenggara) and over much of Australia. So from this point of view, the evolution of similar animals (or plants) is not completely unexpected.

Furthermore, there are giant lizards other than the ora in the southwestern Pacific region (Fig. 6.2). Both the crocodile monitor (*Varanus salvadorii*) and the water monitor (*Varanus salvator*) live here, the former in New Guinea, and the latter through much of Indonesia and the Philippines, although also ranging westward through southeastern Asia as far as northeastern India (Assam and Orissa) and Sri Lanka. Both lizards are reportedly capable of taking small mammals, up to as large as small deer in the case of the water monitor (Bennett 1998). Unfortunately, good data on the prey as well as on the dietary preferences of these lizards simply are not available, so we don't know how often deer are taken (Bennett 1998). Steel (1997) considers the deer "unusual victims." The water monitor reportedly reaches

Fig. 6.3. *The Fijian iguana*
Brachylophus fasciatus. *This
individual was photographed in
the Denver Zoo in 1998.*

lengths of more than 3 meters, but the greatest length that Bennett
could verify was 2.5 meters. The crocodile monitor is a large slender
lizard, with a reported maximum length of 4.75 meters. This would be
as long as individuals of *Megalania,* although the crocodile monitors
are much less massive. However, Bennett (1998) was unable to locate
either the specimen or the zoologist who measured it; the largest
individual he could verify was also 2.5 meters long. He gives considerable detail about the lengths and reported lengths of this lizard and
concludes that quite large monitors may still exist undiscovered in New
Guinea.

Giant lizards also used to live in the Pacific. Fossils of a giant
iguana, currently thought to have been 2–3 meters long, have recently
been found in Fiji (Worthy, Anderson, and Molnar 1999). Fossils of
another (*Brachylophus* sp.) have come from Tonga (Pregill 1989). But
although larger than its surviving relatives (Fig. 6.3), the Tongan iguana
was substantially smaller than the monitors and the Fijian iguana: a
meter long at most. Fossils of both of these animals are still under study,
and the details of their lives and how they evolved may not be known
for some time. However, neither these Pacific islands nor New Guinea
has the same climatic regime as Australia and the Komodo region. New
Guinea has high mountains that precipitate considerable rain, so it is
much wetter than Australia. Fiji and Tonga have a tropical oceanic
climate, without dry seasons. Their climate is more predictable, which
is important, as we shall presently see. So although climate may be a
factor causing similar aspects of evolution in the ora and *Megalania,* it
cannot be a cause of gigantism in lizards in general.

A more plausible cause is simple geography: all of the lands inhabited by the giant lizards we have considered are more or less isolated.
On the map, the Nusa Tenggara and New Guinea (and even to some

extent Australia) seem to be close to Asia, so that one can imagine island-hopping creatures emigrating from Asia to these places. But in fact, few Asian creatures—other than people—have been able to do this, and many of those only with human assistance. The islands of Fiji and Tonga do not even seem close to any major landmass. The large iguanas of these islands evolved in regions not already occupied by other large herbivores—they are both plant eaters—and regions that were difficult for large herbivores to reach. In other words, they evolved in places where there were no competitors: large carnivorous mammals are absent from both the Nusa Tenggara and New Guinea, where the other big monitors dwell. Could this also have happened for monitors in Australia?

It did happen to another unusual—and very rare—lizard, the "panic skink." Skinks are common small lizards of Australia (Fig. 6.4) and almost everywhere else that lizards can live. In wet years, when insects and other small arthropods are abundant, the ground surface in our back garden in Brisbane could be seen to squirm with small skinks. They are sleek, timid little creatures, appearing so streamlined that a common name for one of the larger varieties in eastern Australia is "land mullet." The kinds found in Australia could conceivably create panic if one slipped unseen into your dress, but basically "panic" and "skink" are concepts that simply don't go together.

But in the islands of New Caledonia (northeast of Australia and north of New Zealand), lived a remarkable skink (Fig. 6.5) called *Phoboscincus* (literally, the "panic skink"). This animal is known from only six specimens, all found before 1927. With a total length of about half a meter—as long as five skinks from our Brisbane garden lined up nose to tail—*Phoboscincus* is a big skink. Furthermore, it has strong, prominent jaw muscles, giving it a broad head and sharp, curved, fang-like teeth, not a little like those of *Megalania*, although much smaller. This lizard may never have been observed by zoologists in the wild, and what it ate and how are unknown. But it seems likely to have been the biggest predator on New Caledonia: for a skink, it is an impressive giant.

Once upon a time however, New Caledonia also had an indigenous crocodilian, *Mekosuchus inexpectatus*. It became extinct about two thousand years ago. This was relatively small beast (for a croc), about a meter long, and apparently was a land-dwelling animal, not amphibious like the surviving crocodilians, which fed on mollusks. So it seems not to have been a competitor of *Phoboscincus*. *Phoboscincus* is thought to be a derivative of a skink that made its way to New Caledonia, and once there, evolved into a large (for a skink) predator in the absence of any other predators. There are also fossils of a monitor, possibly the mangrove monitor (*Varanus indicus*), of unknown (but not great) age. This lizard is now found in northern Australia, eastern Indonesia, New Guinea, and western Melanesia. Skink fossils have also been found, but whether or not they might pertain to *Phoboscincus* is unknown, since they apparently remain unstudied. This may be another example of lizards "making good" in the competitive vacuum of the southwestern Pacific, if the giant skink was not contemporaneous

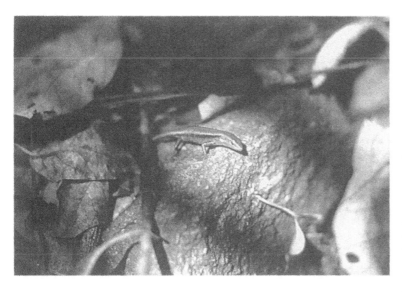

Fig. 6.4. *(top) A common rainforest skink in northern Queensland.*

Fig. 6.5. *(bottom) The "panic skinks" of New Caledonia. The heads of* Phoboscincus bocourti *(left) and* Phoboscincus garneiri *(right), to scale. Both are drawn from preserved specimens collected in the nineteenth century. Scale 1 cm. (Modified from Sadlier 1989.)*

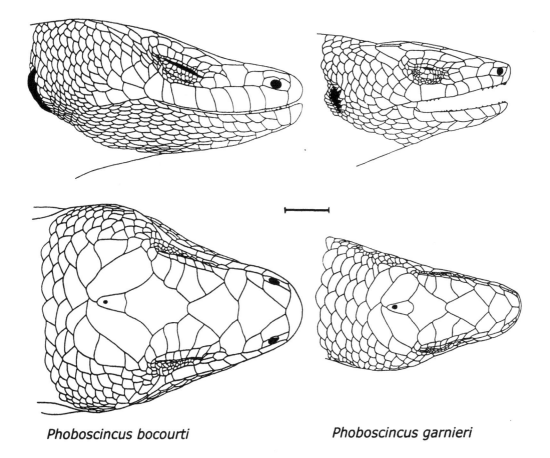

**Phoboscincus bocourti**              **Phoboscincus garnieri**

with the monitor. If they were contemporaneous, there quite possibly was no "competitive vacuum," since the mangrove monitor now hunts a wide variety of prey in a variety of habitats. With *Megalania* and other large monitors in Australia, oras in the Nusa Tenggara, big monitors in New Guinea, giant iguanas in Fiji, large (if not really giant) iguanas at Tonga, and *Phoboscincus* in New Caledonia, one gets the impression that—had humans never appeared here—lizards might have become the major terrestrial vertebrates of this region. One might even add the large land iguanas of the Galapagos to this list.

## The Australian Environmental Lottery

Richard Shine, a herpetologist at the University of Sydney, has argued that the top predators, at least of northern contemporary Australia, are not dingoes but large snakes, particularly pythons. Top predators are those that do not themselves have any creatures preying on them, those at the "top" of the food chain. His argument has yet to be published, although a report on his work appeared in *New Scientist* some years ago (Dayton 1994), and it involves further consideration of climate.

Flannery (1994) observed that Australia is unique in that most of its weather is controlled by a quasi-cyclical change that goes under the hybrid name of El Niño–Southern Oscillation (often abbreviated ENSO). This follows from the predominantly east-west orientation of the continent. In years that are (somewhat arbitrarily) considered normal, the surface waters of the South Pacific off the western coast of South America are cool, but they are warm in Indonesia and off Australia. This creates updrafts in the overlying air and hence rain in the Indonesian region. The elevated air flows eastward and sinks as downdrafts, hence creating dryness at the South American coast. This vertical circulation of air is driven by the warmth of the surface waters in Indonesia and coolness of the waters in the eastern Pacific. It also drives the easterly winds that blow the warm surface waters westward and keep this "pool" of warm water in the western Pacific. But from time to time, the weather changes: the updrafts and downdrafts weaken, thus weakening the easterlies, and the "pool" of warm water extends eastward across the Pacific. The warm waters flow over the cold northward Humboldt Current off the western coast of South America. They evaporate more readily and bring rain to the region, the weather known as El Niño. At this time the western Pacific surface waters cool and inhibit evaporation, bringing drought to Australia. This alternation occurs erratically every two to eight years; its unpredictability governs the unpredictable droughts of Australia. Thus, Australia is subject not only to a mostly uniform climate but also to an unpredictable one.

As I understand it, this led Shine, like Flannery, to argue that the food supply for a large predator in Australia is very uncertain. The uncertainty, as we have just seen, is imposed by the climate, especially via rainfall, and creeps up the food chain from plant to plant eater to meat eater to top predator. If the rains don't fall, the plants don't grow, and the plant eaters die—or at least cease reproducing—and so do those that eat them. Why the climate is so unpredictable is not completely clear—it is the result of the El Niño–Southern Oscillation—but that it *is* unpredictable is indisputable, even at the level of daily forecasts. As I write this in Brisbane, I have just experienced almost two weeks of blatantly incorrect daily forecasting, and this is not uncommon. But to return to the hardships of Australian predators, top predators catch it worst of all, if only because they are usually the largest and thus have the greatest appetites to satisfy. However, ectothermic animals have substantially lower metabolic costs than endotherms. Thus, large snakes can survive from wet season to wet season, no matter how few the prey in between, like those crocs in east Africa that get by feeding on large game once a year. Lions can't do this, and neither—on a smaller scale—can dingoes.

Although there certainly have been large snakes in Australia's past, large mammalian predators are another matter. When Europeans arrived in Australia, the land was distinguished by its lack of large meat-eating mammals. The biggest were dingoes and thylacines, the latter confined to Tasmania, and neither larger than a big dog (in fact smaller than some big dogs). With the attention of zoologists captured by the fact that Australia was inhabited by unusual mammals—the marsupials and the monotremes—the likes of which were rarely seen elsewhere, the absence of big carnivores did not cause much comment. North America has large bears and cats (pumas); South America, large cats (pumas and jaguars); and Europe once had bears. But the really spectacular predators (even if not actually the biggest) were the great cats of Africa and Asia. Although people migrated from Asia to Australia, the big cats did not. And so far as we can tell from the fossil record, Australia never did have large (300+ kg) mammalian predators (although there is a large eagle). Weights and estimated weights of Australian marsupial meat eaters and of placental carnivores from other continents are given in Table 16.

The only reasonably large native predators in Australia were the thylacolions (*Thylacoleo* and kin) and the predatory kangaroos (*Propleopus*). As we have seen with *Megalania*, there has been a tendency to exaggerate sizes. This is true of people in general but is more prevalent and insidious in Australian culture, in which agamid lizards (usually less than 20 cm long) are known as "dragons," koalas as "bears," and thylacines as "tigers." In keeping with this, *Thylacoleo* has been called the "marsupial lion," although it was only as large as a leopard. Murray (1984) estimated the weight, based on two skeletons, as of 75 and 100 kg. Other, but only slightly greater estimates have been made but remain unpublished. A much greater estimate had also been made (Wroe 1999) but was revised downwards to 164 kg (at most), and seemingly very few individuals reached masses this great. Most weighed in between 100 and 130 kg (Wroe et al. 1999). For comparison, this is about the usual mass of a jaguar or leopard.

This was not the first time there was a serious misunderstanding of the nature of thylacolions. It took some time for (some) paleontologists to recognize that thylacolions were meat eaters at all. Owen, who first described *Thylacoleo,* interpreted it as a meat eater. This view was challenged by William Henry Flower, an eminent British mammalogist,

## Table 16
## Weights of marsupial and placental carnivores

| Animal | Weight (kg.) | Weight (lbs.) | Continent | Reference |
|---|---|---|---|---|
| African lion (*Panthera leo*) | 181–227 | 399–501 | Africa, Asia | Walker (1975) |
| spotted hyaena (*Crocuta crocuta*) | 59–82 | 130–181 | Africa | Walker (1975) |
| leopard (*Panthera pardus*) | c. 91 | c. 200 | Africa, Asia | Walker (1975) |
| Indian tiger (*Panthera tigris*) | 227–272 | 501–600 | Asia | Walker (1975) |
| striped hyaena (*Hyaena hyaena*) | 27–54 | 60–119 | Asia | Walker (1975) |
| grey wolf (*Canis lupus*) | 27–79 | 60–174 | Europe, North America | Walker (1975) |
| puma (*Felis concolor*) | 35–105 | 77–231 | North America, South America | Walker (1975) |
| grizzly bear (*Ursus arctos*) | 360 | 794 | North America | Walker (1975) |
| American black bear (*Euarctos americanus*) | 120–150 | 265–331 | North America | Walker (1975) |
| jaguar (*Panthera onca*) | 68–136 | 150–300 | North America, South America | Walker (1975) |
| spectacled bear (*Tremarctos ornatus*) | c. 70–140 | c. 154–310 | South America | Walker (1975) |
| thylacolion (*Thylacoleo carnifex*) | 100–130 e | 220–287 e | Australia | Wroe, et al. (1999) |
| thylacine (*Thylacinus cynocephalus*) | 15–35 e | 33–77 e | Australia | Smith (1981) |
| 'killer kangaroo' (*Propleopus oscillans*) | c. 27–79 e | c. 60–174 e | Australia | Ride, et al. (1997) |

e: Estimate.

in 1868. In 1883 de Vis proposed that *Thylacoleo* was a kind of marsupial hyena, habitually breaking and crushing bones. This view convinced few paleontologists, but it did succeed indirectly in discrediting Owen's view. Robert Broom argued against the herbivorous interpretation of *Thylacoleo* in 1898, pointing out the beast lacked any kind of teeth suitable for grinding or crushing vegetable matter. Charles Anderson of the Australian Museum in Sydney evaluated De Vis's hypothesis some forty years later, in 1929, and concluded that it could not be sustained. At this stage, although there was further inconclusive speculation, it seemed that *Thylacoleo* had discovered a food source completely unknown to paleontologists.

That animals have to eat is not a law of biology but the result of the laws of physics, specifically those of thermodynamics. In eliminating all potential foods, obviously the slow controversy over *Thylacoleo* had seriously overlooked or misrepresented something. In fact, most non-Australian paleontologists were happy to follow Owen, Flower notwithstanding, and regard *Thylacoleo* as a meat eater. But it remained until 1982 for Rod Wells of Flinders University near Adelaide in South Australia and two colleagues to demonstrate that Owen had been right all along. They examined the wear on the teeth of *Thylacoleo* and showed that it was most similar to that on the teeth of other carnivores. They then proceeded to create model *Thylacoleo* molars and subject them to cutting various possible foods. The wear resulting from the cutting of muscles, hide, and ribs of wallabies most closely matched the actual wear. This settled the matter. Wells and his colleagues also concluded that *Thylacoleo* may well have shared the hunting techniques of leopards (Wells, Horton, and Rogers 1982).

The propleopine kangaroos, however, were a surprise. The large modern kangaroos all feed on plants, so there was no warning—so to speak—that some fossil kangaroos were any different. Propleopines were related to the modern rat-kangaroos, or potoroos (potoroids), small beasts who relish a good meal of mushrooms, roots, or insects. The propleopines have been named giant rat-kangaroos—which, because rat-kangaroos were so called because of their relatively small size, is equivalent to being called "giant pygmy kangaroos." But it is not their size that is their arresting feature. One of the front teeth, the third premolar in each lower jaw, is large and flattened, with a round profile and small, sharp transverse ridges. The third premolar of other kangaroos is relatively smaller, enough so that propleopines immediately stood out as unusual. This led to an examination of the other teeth and the discovery that they too were unusual (Archer and Flannery 1985). The incisors had a ventral sheet of enamel, so that tooth wear resulted in self-sharpening teeth. The grinders (molars) were rather low—so that if the animals fed on grass or other tough vegetation, these teeth would wear down relatively rapidly—and they had features (cingula) interpreted to prevent splinters from piercing the gums. Grass or leaves, of course, don't splinter, and these creatures showed no indication that they gnawed wood. Microscopic examination of the teeth showed wear patterns similar to those seen in the teeth of thylacolions, thylacines, and dogs.

In 1997 a team headed by David Ride of the Australian National University in Canberra published a detailed study of *Propleopus*. Ride and his team examined all of the evidence available. As well as the teeth and bones, they looked at the occurrences and abundance of the fossils. They pointed out that *Propleopus* was a rare fossil—consistent with being a carnivore—and has always been found associated with fossils of grazing and browsing kangaroos. Thus, these animals existed in regions in which food—if they were predators—was apparently available. The team also noted that the musky rat-kangaroo (*Hypsiprymnodon*), a living form that dates back at least to the Miocene (about 5–24 million years ago), has an unspecialized digestive tract, and feeds happily on insects as well as nuts and fruits. Since *Hypsiprymnodon* is considered to be representative of early kangaroos, it suggests that the specialized, plant-processing digestive tract of most living kangaroos developed after kangaroos evolved. Thus, there was no obvious bar to the evolution, in the propleopines, of a digestive tract that could handle meat. After carefully studying the structure of the teeth and skull, the team concluded that *Propleopus* was indeed capable of eating meat and suggested that it was probably an opportunistic predator like modern foxes: hunting (or scavenging on) other marsupials when possible, or eating insects and fruits when necessary. Propleopines, although clearly unusual and unexpected, were not particularly large. Ride and his team compared them to wolves in size, or about as big as a thylacine.

The latest thylacoleons were the largest mammalian predators that ever lived in Australia. Reptilian hunters were a different matter. There were large crocodilians: *Pallimnarchus pollens* (now extinct) and the Indopacific croc (*Crocodylus porosus*). The skull of *Pallimnarchus* probably reached about 70 cm in length, so the whole beast was probably at least 5 meters at its biggest. But *Pallimnarchus* had a broad, flattened skull (Figs. 2.28, 2.29), with upwardly directed orbits (sockets for the eyes). This and other features suggest that it was an aquatic ambush predator (Willis and Molnar 1997a). There is no reason to suppose that *Megalania* avoided water—after all, many living monitors frequent watercourses—but the form of its skull suggests that *Pallimnarchus* may have been restricted to water and thus usually not in competition with *Megalania* for prey. But *Pallimnarchus* and the Indopacific croc were not the only crocs about.

*Quinkana* is a Pleistocene Australian ziphodont crocodilian. Ziphodont crocodilians are unusual, interesting, and little-known animals, well worth a brief digression. The term "ziphodont" derives from two Latin words to indicate a tooth resembling a sword. (It was misspelled by the paleontologist who proposed it in the nineteenth century, for the Latin for "sword" is *xiphius,* not *ziphius.*) These crocodilians have flattened, blade-like serrated teeth (Figs. 6.6, 6.7). Some of teeth actually do look (apart from the serrations and with just a little imagination) like miniature sword blades. Some also look like those of carnivorous dinosaurs—for that matter, like *Megalania.* Indeed, until 1977 the teeth of *Quinkana* were mistaken for those of *Megalania.* And this was not the first time such a mistake was made:

Fig. 6.6. Ziphodont crocodylian teeth (AM F.25227 and AM F.25228), found in northeastern Queensland, seen in lateral view. Compare with Fig. 6.7.

Fig. 6.7. The same ziphodont teeth as Fig. 6.6, seen end-on (in mesial or distal view). The flattened form, especially of the larger tooth, may be appreciated. These teeth are also serrate.

ziphodont crocodilian teeth from the Early Cenozoic of Argentina were mistaken for those of theropod dinosaurs when they were first discovered (Simpson 1932). This led to the belief that dinosaurs survived into the Cenozoic in South America—an interesting issue, but too much of a digression for what is, after all, already a digression. We now know that ziphodont teeth evolved repeatedly in crocodilian evolution, and that there were two or three independent lineages of ziphodont crocs in addition to those of Australia (and those like *Dakosaurus* that had ziphodont teeth and lived in the seas during the Mesozoic). These crocodilians (the land-dwelling ones) flourished during the Cretaceous and Early Cenozoic (Palaeogene).

But in Australia they survived well into the Pleistocene, making them the last ziphodont crocs in the world. The Pleistocene animal, *Quinkana fortirostrum,* was a relatively small beast as far as crocs go, probably not much more than three meters overall (Fig. 6.8). However, when Steve Salisbury, then an honors student at the University of New South Wales (Sydney), rummaged through the collections of the Queensland Museum, he was flabbergasted to a find a fragment of *Quinkana* jaw (Fig. 6.9)—from Pliocene deposits—that was much larger than comparable fossils from the Pleistocene (Salisbury et al. 1995). He calculated that the piece might have come from a croc about six meters (give or take a meter) long. This is about as large as the largest Indopacific crocs, the largest Pleistocene crocs, and just possibly (giving a meter) out of the size range of the largest of *Megalania*. This croc was probably the largest known predator of the Australian Pliocene. The largest contemporaneous marsupial predator was the Pliocene species of the leopard-size thylacolion.

The squamates (snakes and lizards) also evolved large predators in Australia, including *Megalania* among lizards, as well as large snakes such as *Wonambi,* the last survivor from an early (Late Cretaceous) lineage of boa-like snakes, the madtsoiids. *Wonambi* was not that large for a snake, not in the same class as *Megalania* among lizards. John Scanlon (another graduate of the University of New South Wales and Australia's only specialist in fossil snakes) responded to the remark that *Wonambi* had a head the size of a shovel with a letter to the magazine *Nature Australia* showing that the shovel must have been of the kind used by small children to build sand castles on the beach. Even so, *Wonambi* was big enough, at about 6 meters long, to give a healthy fright to anyone encountering it—if there was actually anyone around to encounter it before it became extinct. It is likely, although not known for certain, that large pythons ancestral to the living species were also about in Australia during the Pleistocene.

Thus, during the Neogene, Australia had some large reptilian predators: land-dwelling crocodilians perhaps up to six meters long and presumably weighing around 1,000 kg as well as *Megalania, Wonambi,* and other large snakes—not as large as the crocs, but still of a respectable size. There were certainly large crocodilians on the other continents, but these were amphibious or aquatic, not terrestrial, predators.

Compared to those of other continents, Australian mammalian hunters were indisputably small. Weighing in at about 150 kg or less,

Fig. 6.8. The skull, mostly snout, of the Pleistocene ziphodont crocodylian Quinkana fortirostrum (AM F57898) from Cape York Peninsula, Queensland. The skull, such as is preserved, faces to the right. Note the depth of the snout.

Fig. 6.9. The bottom of the right lower jaw (QM F7816) of a giant Pliocene Quinkana, from north Queensland. The jaw is seen from above, and the smooth, oval concavities are the bases of the tooth sockets.

they were comparable to pumas, jaguars, and leopards and smaller than lions, tigers, sabercats, and large bears, not to mention such early —and extinct—mammalian meat eaters as hyaenodonts and amphicyonids. (These last were members of the archaic group of meat eaters known as creodonts, a group that was not ancestral to later carnivorous placentals.) This small size of marsupial meat eaters apparently was not due to any inherent constraint arising from being a marsupial, for some of the marsupial predators of South America such as *Arminiheringia*, *Pharsophorus*, and, perhaps, *Thylacosmilus* (a marsupial analog of the placental sabercats) were fully as large as their placental counterparts.

So if the reason for this generally "small" size was not intrinsic, presumably it was extrinsic—something to do with the environment.

Tim Flannery's hypothesis about the biological productivity of Australia was briefly mentioned in chapter 1. An implication of his ideas is that reptiles are well suited, although probably not adapted, to being the top predators in Australia. The presentation given here follows that of a lecture given by Flannery at the Queensland University of Technology (Brisbane) in May of 1996, which presented a concise summary of the thesis of his 1994 book. Productivity in the biological sense is the capacity to support large numbers or a large diversity of organisms—plants and animals are the organisms usually considered, but only because few ecologists are seriously interested in fungi and bacteria. Productivity obviously depends on the availability of the resources required for living things. These include sunlight, water, and nutrients. As we all know, plants are at the root of this process because they are the major organisms that can subsist on light, water, and inorganic compounds (minerals) alone.

Examining the supply of these prerequisites for productivity in Australia, Flannery noted that sunlight is actually the only one that is abundant. Australia is obviously a desert continent, where large areas receive little rain—and that unpredictably and often very sporadically (an observation already made before 1923 by Australian geographer Griffith Taylor). But Flannery noticed that Australia is also poor in inorganic nutrients. We have often heard that the soils beneath tropical rainforests are poor because all of the minerals are tied up in the vegetation. But the minerals were there in the first place and were extracted by the plants. In Australia, they are no longer there "in the first place." It is not that the soils of Australia are inherently different from those of other lands, but that they are old. They have been deeply leached by rains falling since the middle of the Cretaceous, about 90 million years ago (at least in places like the west-central plains of Queensland). Over the millennia and more, the rain waters dissolved the soil minerals and carried them off to the surrounding seas—where doubtless they were welcomed by the marine life. This process also happens elsewhere, on other continents, but—Flannery realized—the difference was that these continents also experienced three geological processes that created new fertile soils: mountain-building, glaciation, and volcanism.

Being centrally located in a tectonic plate—as we have already noted—Australia was removed from those areas where serious mountain-building was taking place. The major mountain system of Australia, the Great Dividing Range lying only 50 to 400 kilometers inland of the east coast, is approximately 90 million years old, thus dating from 25 million years before the dinosaurs became extinct. At about a kilometer high (about 3,300 feet) for most of the range, these mountains would barely be considered foothills in Colorado or Switzerland—or New Zealand, for that matter. As such, they are not exposed to same degree of erosion as the much higher mountains elsewhere, and so produce less soil. This is not to say that they produce none at all: as well as parts of Victoria and Tasmania, the Darling Downs of southeastern Queensland have good soils, produced by this erosion (as well as volcanism), that not only supports agricultural communities but also preserves fossils of Pleistocene vertebrates like *Megalania.*

The major grain-producing regions of the world, the Great Plains of North America and the steppes of the Ukraine, owe their productivity to their soils. And they owe this ultimately to the cold of the Pleistocene, to the grinding of rock—slow, but exceedingly fine—by the advancing ice sheets of the Northern Hemisphere. As we saw in chapter 1, this happened repeatedly in the north. But the continents of the Southern Hemisphere were either not close enough (Australia) or too close (Antarctica) to the South Pole for this to occur. There were, indeed, glaciers in Australia, but compared to those of North America and Europe, they were small things, the largest about 50 square kilometers, corresponding in their icy way to the small "mountains" of Australia. These miniature ice sheets contributed little to soil production. And much of what they did contribute was lost. The arid, windier climate of the Australian Pleistocene that we mentioned in chapter 2 blew the fine grains out into the South Pacific Ocean to sink to the bottom, where they remain to this day.

Just as there is little mountain-building in the middle of a tectonic plate, so there is also little volcanism. The so-called Australian volcanoes are in New Zealand and New Guinea and their associated islands. But there is some volcanism in Australia, mostly along the eastern coast. The 23-million-year-old (Miocene) Tweed Volcano just across the border of New South Wales south of Brisbane has now eroded into a crater almost 25 km wide. (The name "Tweed Volcano" refers to the Miocene mountain; its remaining remnants include the Lamington Plateau and the Tweed and Nightcap Ranges.) This and other Miocene volcanoes account in part for the rich soils of the Darling Downs mentioned above, and even more recent volcanoes created the rich farmland of western Victoria—but these are all "oases" in a desert of lean, barren soils.

So Flannery's hypothesis invokes, first, the long-term leaching and loss of fertile soils in Australia, and second, the absence of any significant mechanisms for producing new soils. His argument then proceeds to claim that poor soils promote the evolution of a diversity of different plants—as are found in the southwestern corner of the continent—whereas richer soils promote less diversity. I am not convinced by this, or at least by the second part. But this is also peripheral, or even irrelevant, to the stream of argument linking plants to carnivores. Diverse the plants might be (and are in the southwest), but productive they are not. This is not due to anything intrinsic in Australian vegetation—in Victoria and Tasmania, where the soils are fertile, eucalypts can grow to be over 120 meters (400 feet) tall—but because of the wide extent of the infertile soils, and also because of the third factor that we have yet to consider: water.

As discussed previously, the surface waters of the South Pacific vary in temperature. A large "pool" of warm water often forms in the west, near Australia and New Guinea, but sometimes much further east. The position of this "pool" governs the rainfall and general storminess of the bordering lands. It also governs Australia's weather, and it is unpredictable—even, to some extent, to a meteorologist with a super-computer. If the rain is unpredictable, the availability of water is unpredictable. In chapter 2 we discussed the position of Australia in relation to the subtropical convergence and its implications. The significant one here is that the air over Australia tends to be very dry, so that ponds and lakes dry out over the millennia unless supported by adequate and reliable rainfall. And that is just what they did not have once the last glaciation ceased (and, presumably, during the previous interglacials as well). There are large numbers of large and small lakes in the outback, but all of them are dry. They will not hold water again until glacial conditions return or Australia creeps north into the tropics.

The response of many plants (and not a few animals) to this is to take maximal advantage of the rains when they do show up. The unpredictability of the rain, and hence availability of food, contrasts with the predictability of seasons in the Northern Hemisphere. Tim illustrates this with grizzly bears, who avoid the nastiness and hardship of winter by hibernating. They do this certain in the (evolutionary) "expectation" that with the coming of spring—with increased sunlight and available water, supported by the good soils—there will be more than adequate vegetation to support both the grizzlies and the plant eaters that grizzlies fancy from time to time. The availability of sunlight and water is governed by a reliable yearly cycle (the soil, of course, is there all the time). In Australia, what passes for a cycle is governed by the Southern Oscillation, and it is neither yearly nor reliable. Those animals that do "hibernate" (aestivate, actually), like the water-holding frogs (*Cyclorana platycephalus*), must be able to survive for a long but unknown period of time before the water returns. This ability favors metabolically "cheap," ectothermic animals, like the water-holding frogs. Since they "spend" their energy more slowly than mammals and birds, they can spend it for longer.

So it is not really the low productivity of the plants that is important here; it is that this productivity is unreliable. If the supply of plants is unreliable for plant eaters, we expect that the supply of herbivores will be unreliable for carnivores. And we are not disappointed. The major placental herbivores in Australia—in no small part because before the coming of Europeans they were almost the only placental herbivores in Australia—are rats. And in Australia, with rats come rat plagues. In good years (after the rains) there may be 2,500 kg of rats per square kilometer. In poor years, it is 300 kg of rats per square kilometer, a fall of almost 90 percent. This too favors the metabolically "cheap," ectothermic predators—reptiles—who can thrive on about 10 percent of the food required for a mammalian predator of the same mass. Studying these pythons, Shine found that in spite of the unpredictability, the rats could still support 400 kg of python per square kilometer. Much more python biomass was supported than biomass of dingoes and thylacines. Even the biomass of predators in east Africa, where the rainfall (and hence vegetation) is reliable, is only about one-tenth as much (all figures from Dayton 1994).

Given this interpretation, it should not be surprising that very large reptilian predators evolved in Australia. *Megalania,* if our approach is valid, would have combined the benefits of a varanid circulatory and respiratory system for distributing oxygen, as well as the benefits of varanid heart function, with the advantage of an energetically "cheap" ectothermic physiology that was enhanced by a still lower metabolic rate as its large size increased. This combination would have been just the thing for a land where prey was plentiful but only in unpredictable fits and starts, so that the energetically more "expensive" mammalian predators were often at a disadvantage. We might suggest that the success of *Megalania* was due to a combination of its unusual (for lizards) circulatory system, its effective digestion, and the absence of large marsupial predators. In other words, it had low "running costs" and lacked serious, sophisticated competition from mammals.

## Productivity and Monitors in Eastern Indonesia

Although I do not accept it in its entirety, Flannery's scenario for the evolution of climate, soils, plants, and animals in Australia is unquestionably a remarkable achievement, providing the basis for a general theory of how life evolved in arid, unpredictable Australia. Conceiving this scenario was made easier by the geographical isolation of Australia through most of the Cenozoic. With suitable modification, it should provide an enlightening framework for understanding any ecologies of this part of the world. If so, it should be possible to assess how well it applies to the Nusa Tenggara.

Like Australia, Flores is longer from east to west than from north to south. However, Flores—like the rest of the Nusa Tenggara—is too small to extend over more than one climatic zone. So it does have a similarly uniform climate, but this is due to its size, not its geographical orientation. But unlike Australia, Komodo and Flores are in the tropical region at the inter-tropical convergence. Here air near the ground, converging from the north and south, tends to get heated and rise, unlike the subtropical convergence where air at altitude tends to converge, cool, and sink toward the ground. The rising air does eventually cool, and its moisture precipitates, generating the intense rains and storms of the Tropics.

This nicely accounts for the rain forests of western Indonesia and Indochina, but how, then, do Komodo and its vicinity come to be dry? The answer is as much in the geography as in the climate. The December to March monsoons generally move from northwest to southeast. Thus, they encounter the mountains of Kalimantan, Java, and Sulawesi, where most of their rain falls before approaching Komodo and Flores. The June monsoons, which move in the other direction, originate near or over Australia, where there is little water to evaporate in the first place. In fact, most of the weather of these islands, from April to November, is "carried" by air from Australia moving over the islands toward the subtropical convergence. The result, dryness, is the same, but the cause is different: climate in Australia, geography in the Nusa Tenggara. The weather of the Komodo region is no less certain than weather anywhere else, but the climatic unpredictability of Australia is absent here: the weather is similar, but the climate is not the same.

Nor are the soils here as poor as those of Australia, for it is here that the volcanism, largely absent from Australia, occurs. So the similarity is not in the causes, but in the results. It is the aridity alone that limits the productivity here, as opposed to both aridity and poor soils in Australia. Additionally, geography plays another role in this part of Indonesia. Many of the islands inhabited by oras are small. They can support small populations of prey animals—deer and boar—but are probably too small to support, in turn, the likes of tigers and leopards. The placental carnivores, large or small, do not range east of Bali, the edge of the Asian continental shelf.

## A Competitive Vacuum?

The other giant lizards of the southwestern Pacific (New Guinea and Fiji) occur in quite different habitats, and those of Fiji and Tonga were, after all, herbivores. The sole common factor for all these lizards seems to be geographical isolation. Although relatively large, New Guinea was never part of any larger landmass except Australia (which by that time was also isolated). Tigers and deer never got as far east as New Guinea and New Caledonia: the monitors had it to themselves (except for the crocs and other lizards). Fiji and Tonga were farther away from Asia, and even monitors did not make it that far.

This all makes a good story. Large lizards got large because—to phrase it metaphorically—they seized opportunities that other creatures could not. This, in turn, is because these other creatures had some limitations that the lizards did not have—either in getting to remote places, or in surviving in climates that were more like lotteries than like the usual rhythm of the seasons. We have pointed out that Australia lacked large marsupial predators, but what about the others, the reptilian ones? They were certainly there. So if competition had something to do with the success of *Megalania*, we would expect that megalanian fossils usually are not found together with those of crocodilians, especially *Quinkana*. This does seem to be the case. *Megalania* seems to have enjoyed the eastern Darling Downs region, or at least its fossils are more readily found there than elsewhere (which doesn't mean that one would call them common). Crocodilian fossils are notably rare in this region: they do occur, but there are literally only a handful, and they are a cause of excitement when found. Ziphodont crocodilian teeth—although nowhere common fossils—are completely absent here. Although we must always remember that the fossil record is biased, it looks as if *Megalania* was the major reptilian predator of this region. For the Pliocene, the story is not quite so clear, but the fossils of the giant *Quinkana* come from a region in north Queensland where *Megalania* was rare or absent. Although crocs were seemingly not rare at Chinchilla (in the south), they were amphibious crocs, and ziphodont teeth are again rare. The evidence is consistent with the idea that *Megalania* inhabited regions where crocodilians were either very rare or absent or where only amphibious crocs lived.

Although I think this is an important part of the story, it clearly is not the whole story. The remains of the other extinct giant monitor, *Varanus sivalensis,* were found in the nineteenth century and hence lack detailed data from which their age may be derived. They are generally regarded as having come from the Pliocene Siwalik beds. These beds were laid down in a region inhabited by elephants (including deinotheres), horses, pigs, hippos, and deer, not to mention a variety of rodents and primates. Among the meat eaters were amphicyonids, bears, hyenas, and sabercats—clearly not the equivalent of the small carnivorous mammals found in Australia at this time. The few isolated fragments of *Varanus sivalensis* seem to show that giant monitors could survive in regions that were not isolated and where large and presumably "sophisticated" mammalian predators were significant. The Trinil and Kedungbrubus oras were also found in faunae with large, advanced placental carnivores (e.g., tigers in the Trinil and hyenas in the Kedungbrubus). So although the oras of the Komodo region and *Megalania* seem to have lived and evolved (in the case of *Megalania,* but quite possibly not the ora) in regions where the combination of efficient varanid physiology and low productivity made them very effective predators and competitors, this is not a general theory for varanid success—not even for the success of the giant monitors.

In would be in no way surprising if there were no single factor that could be adduced to account for the success of the giant varanids—although it would be satisfying if there were. The answer may well be in the study by Losos and Greene (1988) mentioned in the previous chapter. They looked at diet in a variety of monitors and its relationships to their evolution and concluded that the view that monitors are generally predators of large vertebrates, like cats or dogs are, is not correct. They believe that the success of monitors depends on their physiology and their ability to eat a wide variety of foods. They suggest that the archetypal monitor had food preferences much like those of foxes. Presumably, it was their catholic diet that underwrote their success by permitting them to switch to whatever food was readily available. Although not so dramatic as the hypothesis of environmental limitations on productivity, it does invoke an issue close to the heart of all animals—getting fed.

*One measures a circle*
*beginning anywhere.*

—C. H. FORT, 1931

# 7. Summary:
# How a Dragon Came to Be

During the Pleistocene, when *Megalania* flourished, the world was different. We easily assume that the past was much like today—and in some ways, and at some times, it was. At some times, however, it wasn't. It is not possible to have an understanding of an animal or how it lived without knowing the world in which it lived. Understanding or appreciating the world of an animal much smaller (or larger) than us is difficult enough today, and hence it is even more so when that animal is extinct and so are some possibly very significant aspects of its world.

The climate of the Pleistocene world of *Megalania* shifted between being as warm or warmer than it is today, and somewhat cooler. The "cooler" periods were not greatly colder than today, but enough so that in the polar and near-polar regions the snow did not melt entirely over the summer and eventually came to form ice sheets more than 2 km (about 1.2 miles) thick. Once formed, this ice reflected the sunlight and so helped to maintain the cooler climatic regime. The water frozen into the ice was withdrawn from the oceans, and the sea level fell to about 50 meters (about 165 feet) below that of today. At some times, it was much lower. While the ice accumulated in the northern regions—and Antarctica—some regions became more arid. This shift is more relevant to a dry continent such as Australia—and to its lizards.

This kind of climate vanished from time to time, and a climate apparently much like that of the present took its place. These changes seem to have been rather abrupt, occurring perhaps in as little as twenty years. The changes from glacial to interglacial times were marked by catastrophic floods resulting both from the escape of dammed meltwater near the ice sheets and from the relatively rapid rise in sea level. Such changes must have placed many organisms, both animals and plants, under stress—not so much from the catastrophic events as from the alterations in climate and their results on the plant communities. These changes would have reverberated up the food chain to affect even creatures like *Megalania*. This much we can guess, but so far the fossil record is too poorly known to reveal their effects in Australia.

During the glacials some parts of the world, such as northeastern Siberia, seem to have had habitats, or ecosystems, that no longer exist anywhere. That area was probably a vast open grassland that remained well below freezing for the entire winter, during which the sun did not rise because of its arctic location. In the summer, when the sun did not set, the weather would have been warm, over 30°C (85°F), perhaps even warmer, and clear. Given the apparently complete disappearance of some environments, we should not make assumptions about the environment of *Megalania*. Not only do we lack good data about the paleoecology of Australia, but even "good" data may not be enough if that ecosystem was unlike anything that now exists.

Geography is a factor in the evolutionary history of animals and plants. The periodic rises and falls of sea level during the Pleistocene repeatedly changed the connections between lands. With low sea stands, continents were joined, and islands were enlarged or even created; with high stands, continents were separated, and islands were reduced or submerged entirely. During low stands of the sea, Australia considerably increased in area and was continuous with New Guinea (Fig. 2.20).

In Australia, the fossil record for terrestrial animals in effect begins with the Miocene. There are enough fossils to sketch out the preceding history back to the beginning of the Mesozoic, but they provide merely a few glimpses, not a view. Already in the Early Triassic, about 240 million years ago, when Australia was part of the southern supercontinent of Gondwanaland, it may have had a distinct fauna. This was different, not in the kinds of animals present, but in that the temnospondyl amphibians formed a larger proportion of the whole than elsewhere (Fig. 2.10). In the Cretaceous, Australia had both unusual endemic dinosaurs and relict—"living fossil"—temnospondyl amphibians. Although many of these were creatures very similar to those that lived elsewhere, at least some components of the fauna were clearly distinct. It seems that Australia had "predeveloped" its own unique fauna even before it became the isolated island-continent. The entry of the monitors into Australia, presumably near the opening of Miocene times, brought them into a land with a fauna different from any they had previously encountered. Judging from the results—a greater variety of monitors in Australia than anywhere else—it suited them.

The Cenozoic history of Australia itself is the story of a slow journey north from the south polar regions toward the equator. As Australia crept northward into a warmer but drier climate, the climate of the earth as a whole cooled. The result for Australia was that the climate and the land dried out. In the vast central parts of the continent, rain forests contracted toward the coasts and were replaced by open, drier woodlands, then grasslands, and, finally, desert. Not only did the climate become drier, but it also became more erratic, and the rains less predictable. However the early Australian monitors coped with the forests, they seemingly did well regardless of these changes in climate and their effects on the plant communities.

Either by chance—because their ancestors almost alone had inhabited that region of Gondwanaland that became Australia—or because their reproductive strategy was more suited to the increasingly erratic climate, marsupials had become the dominant mammals. But the unpredictable climate also favored those creatures that could survive the droughts because they needed less food per kilogram body weight—reptiles. So Australia came to have an extensive snake and lizard (and monitor) fauna.

Not surprisingly, and in accord with the principle of parsimony ("Ockham's razor"), the early attempts to understand Australian vertebrate fossils interpreted them in terms of what was known at the time. Thus, at least some Australian fossil animals were initially interpreted as Australian representatives of animals previously found elsewhere, though increasing knowledge of both Australian and other fossil animals made this an untenable view. *Megalania* was initially interpreted in light of a view (perhaps tacit) of Australia as an isolated remnant of Mesozoic Asia. This was seemingly related to regarding more "archaic" animals, such as reptiles, as more "robust" than more sophisticated forms like birds and mammals. These robust, "archaic" forms could thus better survive the rigors of "island-hopping" along the islands and rafting across the seas of southeast Asia to ultimately reach Australia. Again, further finds of fossils and further study of them, both outside and within Australia, revealed that this view, too, was incorrect. The current view is that Australia comprises a mixture of forms, some deriving from when it was still part of Gondwanaland, and others—the monitors included—having moved from island to island, from Asia to Australia.

Similarly, the attempts to understand new and novel extinct animals from Australia like *Megalania* inevitably led to confusion—although in this case the confusion was largely limited to Owen. The history of the understanding of *Megalania* is largely a history of the discussion of whether or not the genus is valid. Little in the way of new material adding substantially to our understanding of the beast was discovered between the late nineteenth century and 1990. New discoveries, however, have stimulated new interest in paleobiological interpretation. This is not to say that interest in paleobiology had not previously existed, but it was largely limited to scientists of continental Europe and the United States.

The fossils of *Megalania* seem to show that it lived in the eastern part of Australia and did not inhabit the west, Tasmania, or New Guinea. The reasons for this remain unknown, although it is possible that *Megalania* did actually live in these places and that its fossils remain yet to be found.

The fossil record of lizards in Australia essentially begins in the Miocene, although a few older (but fragmentary) fossils are known. All of the major groups now living in Australia were already present by that time. Thus, if monitors came to Australia from Asia, as the available evidence indicates that they did, they presumably came along the more extended archipelago that then existed, since Australia was then further from southeast Asia. For a group of lizards that has colonized as far as New Caledonia and the islands of western Melanesia east of New Guinea, this is not unexpected.

By and large (although not entirely), it was the larger of the Australian reptiles that became extinct in the Pleistocene. This is consistent with the extinction of the larger mammals and birds also at this time. But we must remember that this conclusion may be misleading because the smaller forms are very poorly known as fossils (and to a lesser extent this is true for mammals and birds as well). Australia seems to have been host to a variety of relict Gondwanan reptiles that did not survive the Pleistocene—a kind of zoo for past kinds of animals, now disappointingly forever closed.

The fossil record of monitors and their ancient kin indicates that they evolved from their anguid ancestors on one of the northern Laurasian continents, probably what is now Europe or North America. This conclusion is based on the discovery of a potential ancestor of the platynotans, *Parviraptor,* in Europe and North America, but also on the discovery of fossils of these forms from the Cretaceous of Asia and North America and their absence from southern continents (Africa and South America). The genus *Varanus* itself appears in the fossil record about 25 million years ago, just before the beginning of the Miocene. Shortly after this time (geologically speaking), it seems to have been widely spread in Eurasia and to have reached both Africa and Australia—implying that it was also in the southeastern Asian region. Fossil specimens of monitors seem in general to be preferentially composed of vertebrae, and monitors are rare fossils.

The fossil record of Indonesia dates back probably less than four million years (Early Pliocene). Even so, fossils of varanids have been recovered from several of the western islands, and it is clear that large monitors, presumably the ora (*Varanus komodoensis*), lived on Flores —and possibly Timor and Java—early in the Pleistocene.

Biochemical studies on monitors indicate that most of the Australian monitors form a closely related group. One species, however, is closely related to the ora and the crocodile monitor (of New Guinea), forming a second group, and the other Asian monitors form a third closely related group. The African monitors are not closely related to the others. This is consistent with the evidence of the fossils that monitors originated in Asia or Eurasia and separate lineages moved into Africa and southeast Asia. One or two of the Asian lineages moved into southeastern Asia and on to Australia. Once in Australia, one lineage diversified into all of the living Australian monitors except the lace monitor. The other produced the lace monitor, the crocodile monitor, and the ora. Unfortunately, there is yet no indication to which of these lineages *Megalania* is related. Relating it to the ora, although tempting on the basis of their large sizes, is premature.

*Megalania* dates back to early Pliocene times, around 4–4.5 million years ago. Fossils of the perentie, Gould's goanna, the lace monitor, and possibly the crocodile monitor come from the Pleistocene (or perhaps earlier) and show that a variety of large monitors has inhabited Australia for the past one or more million years. On Komodo, the ora excludes other large monitors, but the Australian data is still too incomplete to show whether the smaller (but still large) monitors lived together with, or separate from, *Megalania*.

There have been many estimates of the size of *Megalania* but no explanations of how the estimates were arrived at. My own measurements indicate that Hecht's estimates are the most plausible. These indicate that the largest known individual of *Megalania* was about seven meters (about 23 feet) in total length—but this assumes that the tail was about as long as the body, not, as Hecht assumed, that the tail was about half as long as the body (snout–vent length). Hecht's assumption here is unlikely, since there is no evidence that any monitor, living or fossil, had a tail significantly shorter than the snout–vent length. As in oras, the largest individuals might well have been males.

Estimates of weight—like those of length—vary widely. Again, those of Hecht seem plausible, or even a little high. I estimate that a 2.3-meter *Megalania* may have weighed about 185 kg (a little less than 410 lbs.). The estimates of weight were derived by several methods, the most reliable of which seems to be the revised equation of Blob.

Based on comparison with oras, *Megalania* may have been ambush hunters, but, also like oras, sometimes giving a short chase. Septic bacteria inhabiting the mouth may have played a role in obtaining prey, and *Megalania* may have followed more or less habitual routes when foraging. But the most important features relevant to hunting were physiological. The stamina of living monitors is at least in part due to: (1) a technique of breathing that allows monitors to force air into the lungs while moving quickly; (2) blood that retains its capacity to carry oxygen longer than that of other lizards; and (3) an effective heart that prevents mixing of oxygenated and deoxygenated blood in the circulation and supplies blood to the muscles of the body at higher pressure than the blood flowing to the lungs. In addition to these is the efficiency of feeding and digestion seen in the ora and other living monitors. Because of its large body size, *Megalania* may have been more independent of changes in temperature than smaller monitors, an effect also seen in oras. These features, taken together as they presumably were in *Megalania,* suggest that it was a formidable predator.

Unlike the skulls of smaller monitors, the skull of *Megalania* lacked internal joints (i.e., cranial kinesis), presumably to strengthen it against the weight of the head. In living monitors, these joints seem to aid in the swallowing of food by increasing the extent to which the mouth can be opened—often relatively large pieces are bolted. The head of *Megalania* may well have been large enough that this was unnecessary. The curved, serrated teeth of *Megalania* probably cut flesh and hide more easily than if they had been straight and lacked serrations. They are similar to those of other meat-eating reptiles, including carnivorous dinosaurs. Like living monitors (and other meat eaters), *Megalania* probably both hunted live prey and scavenged carcasses.

*Megalania* is assumed to have been territorial, like oras. Estimates of the territory size of an individual *Megalania* are about 5 to about 30 square kilometers. Doubtless the range varied depending on the numbers of prey available. Estimating the numbers of *Megalania* relative to those of contemporaneous fossil crocodilians suggests that *Megalania* was substantially less abundant (in Queensland as a whole) than crocodilians. On the Darling Downs, where a large proportion of fossils of *Megalania* have been found, crocodilian fossils are quite rare. This suggests that *Megalania* and crocodiles did not live together. Rough estimates of the amount of food required by an adult *Megalania* weighing 460 kg, based on data for living monitors and combined with published estimates of the weight of various Pleistocene Australian marsupials, suggest that *Megalania* could have survived on the consumption of two diprotodonts or ten to twelve large kangaroos (sthenurines) per year.

In the early 1990s the discovery of bones of the skull roof showed that *Megalania* had a low cranial crest. Cranial crests are also found in Cretaceous "proto-varanids," but this is the first instance in a Cenozoic varanid. This is interpreted as having been a sexually dimorphic feature used in fighting or displaying for a mate.

Based on data for Rosenberg's goanna (*Varanus rosenbergi*), a 3.5-meter-long (about 11 feet 6 inches) female *Megalania* would lay about 25 eggs that would require incubation for about 280 days. Such figures are not obviously implausible, but if true suggest that, unlike behavior reported for oras, female *Megalania* would not have refrained from eating through the incubation period. Rosenberg's goanna requires an additional 32 percent of food, minimally, to lay eggs. Given the modest food requirements estimated for *Megalania*, this would not seem usually to have constituted any hardship for females.

*Megalania* is one strand of, if not a fabric, at least a woven cord. The evolution of *Megalania* is not completely separable from that of monitors in general. Monitors may well be unusual for lizards in the sophistication of some of their physiological features and processes, but they are unusual among tetrapods in the conservatism of their bodily form. Given the current evidence, it seems plausible that the primordial varanid was a moderately large animal and that different lineages have become both smaller and larger still. Since *Megalania* is the largest of these, it is the evolution of large size that is of interest here. There are many potential factors for this, and unfortunately we cannot determine which are responsible. However, it does seem that the advantages of thermal independence following from the large size during the cool weather of the Pleistocene can be eliminated, since *Megalania* was already large early in the Pliocene. Sexual selection (and competition for mates, if these actually act independently) and advantages in hunting are possible factors. A constraint acting to prevent their becoming even larger may have been the sprawling posture and gait.

Large monitors in Australia may have held a competitive advantage over marsupial meat eaters—which seem never to have produced animals as large as the placental carnivores of other lands—in having been ectothermic. Ectothermic predators generally require about one-tenth as much food for any given period as endothermic ones. They could thus more easily "weather" the unpredictable droughts of the Australian climate in which prey animals may have become rare. The arid climate of Komodo and associated islands may have given the ora a competitive edge over endothermic predators—but there is no evidence that endothermic predators actually reached the region before the arrival of humans.

Large lizards of the southern Pacific region—both carnivorous and herbivorous—seem to have evolved in a region where there was little competition (although competition was not necessarily totally absent). But these factors probably cannot account for the evolution of large size in all giant monitors. *Varanus sivalensis* reached a body size probably of about three meters although living on the Asian mainland with a full complement of placental mammalian meat eaters.

Though being able to reach and live in places that potential competitors could not reach is probably a significant factor in the evolution of giant varanids, the wide variety of habitats that they can live in, the wide variety of potential foods, and their advanced physiological features together with their "cheap" ectothermic metabolism would probably provide a more general explanation for their success. But the specific factors acting in each specific case (each species) may well be different.

*Megalania*, and its fellow creatures of prehistoric Australia, should remind us that the world can hold more biological diversity and richness then we might often think. The existence of giant lizards, now restricted to a few Indonesian isles, was unexpected from the Northern Hemisphere perspective of our nineteenth-century intellectual ancestors. That New Zealand, before the coming of people, had a land-dwelling fauna in which the large creatures were all birds was equally unanticipated. The difficulty, then, was separating the effects of historical occurrences—often viewed as chance occurrences—from those determined by ecological and physiological mechanisms, a difficulty that is still not always satisfactorily resolved.

It may seem out of place at the end of a whole book about *Megalania* to point out that our understanding of this creature is at a rudimentary level—although the perceptive reader has doubtless reached that conclusion already. But research on *Megalania* continues. The donation of further material to the Queensland Museum in the 1990s made possible new studies carried out while this book was being written and still unpublished. One of these is a study of the histology of *Megalania* conducted by Vivian de Buffrénil and Armand de Ricqlès in Paris. One hesitates to "sacrifice" fossil bones to be cut up for histological studies if the bones are rare and hard to come by, but the recent finds of *Megalania* have made such studies feasible.

These studies show that the cortical (surficial) bone structure is similar to that of other monitors but that some bone also shows remodeling of the structure known as Haversian tissue. These kinds of structures are generally found in (large) mammals but are also seen in dinosaurs. Their presence has been used as an argument for endothermy, but they seem more likely related to large size. Certainly no one (so far!) has suggested that *Megalania* was endothermic. But with a short neck and stout legs that would have restricted heat loss, large individuals of *Megalania* may well have been able to maintain body temperatures above those of their environment.

De Buffrénil and de Ricqlès suggest that the period of rapid growth characteristic of the young of large living monitors may also have occurred in *Megalania*. In addition, its large size suggests a longer life span. This increased period of rapid growth, together with a generally longer life, would account for the large size of this creature.

Greg Erickson in Florida, working together with de Buffrénil, de Ricqlès, and Mark Bayless, examined the histology of the small worm-like bones of the skin (osteoderms). They found that these bones showed annual growth rings that indicate that individuals grew rapidly for the first twelve years and then slowed—but did not stop growing altogether. Twelve years is longer than the period of rapid growth in any living monitor and indicates that the conjecture of de Buffrénil and de Ricqlès that *Megalania* had relatively long lives is likely. The Alderbaran Creek individual died at age 16, and was still growing.

These results have further implications than just illuminating the lives of *Megalania*. De Buffrénil and de Ricqlès have pointed out that increase in body size in such active creatures as monitors would result in decreased heat loss (because of the relationship between surface area and volume). This, combined with periods of increased metabolic rate, would tend to produce a relatively high and relatively constant body temperature (gigantothermy). They suggest that instead of a single evolutionary route to endothermy, there may be several, with different paths being taken by the ancestors of mammals and those of birds (i.e., dinosaurs). Thus, studies on *Megalania* may help us to understand not only this great lizard but also the spectrum of evolutionary developments over the history of tetrapods.

Abler, W. L. 1992. The serrated teeth of tyrannosaurid dinosaurs, and biting structures in other animals. *Paleobiology* 18: 161–183.

Aksu, A. E., R. N. Hiscott, P. J. Mudie, A. Rochon, M. A. Kaminski, T. Abajano, and D. Yasar. 2002. Persistent Holocene outflow from the Black Sea to the eastern Mediterranean contradicts Noah's Flood Hypothesis. *GSA Today,* 12:4-10.

Allen, J., and J. F. O'Connell, eds. 1995. *Transitions. Antiquity* 69, spec. no. 265.

Anderson, C. 1929. Palaeontological notes no. 1: *Macropus titan* Owen and *Thylacoleo carnifex* Owen. *Records of the Australian Museum* 17: 35–45.

Anderson, J., A. Hall-Martin, and D. A. Russell. 1985. Long-bone circumference and weight in mammals, birds and dinosaurs. *Journal of Zoology* 207: 53–61.

Andersson, M. 1994. *Sexual Selection.* Princeton: Princeton University Press. 599 pp.

Aplin, K. P., J. M. Pasveer, and W. E. Boles. 1999. Late Quaternary vertebrates form the Bird's Head Peninsula, Irian Jaya, Indonesia, including descriptions of two previously unknown marsupial species. In A. Baynes and J. A. Long, eds., *Papers in Vertebrate Palaeontology,* pp. 351–387. *Records of the Western Australian Museum,* Supplement, 57.

Archer, M., and T. F. Flannery. 1985. Revision of the extinct gigantic rat kangaroos (Potoroidae: Marsupialia), with description of a new Miocene genus and species and a new Pleistocene species of *Propleopus. Journal of Paleontology* 59: 1331–1349.

Archer, M., S. J. Hand, and H. Godthelp. 1991. *Riversleigh.* Sydney: Reed Books. 264 pp.

Arvidsson, R. 1996. Fennoscandian earthquakes: Whole crustal rupturing related to postglacial rebound. *Science* 274: 744–746.

Auffenberg, W. 1972. Komodo dragons. *Natural History* 81, no. 4: 52–59.

———. 1980. The herpetofauna of Komodo, with notes on adjacent areas. *Bulletin of the Florida State Museum, Biological Sciences* 25: 39–156.

———. 1981. *The Behavioral Ecology of the Komodo Monitor.* Gainesville: University Presses of Florida. 406 pp.

———. 1988. *Grays' Monitor Lizard.* Gainesville: University Presses of Florida. 419 pp.

Avise, J. C., and D. Walker. 1998. Pleistocene phylogenetic effects on avian populations and the speciation process. *Philosophical Transactions of the Royal Society of London*, Ser. B, 265: 457–463.

Baker, V. R., G. Benito, and A. N. Rudoy. 1993. Paleohydrology of Late Pleistocene superflooding, Altay Mountains, Siberia. *Science* 259: 348–350.

Bakker, R. T. 1986. *The Dinosaur Heresies*. New York: William Morrow and Co. 484 pp.

Balouet, J.-C., and E. Buffetaut. 1987. *Mekosuchus inexpectatus*, n. g., n. sp., Crocodilien nouveau de l'Holocène de Nouvelle Calédonie. *Comptes Rendus de l'Academie des Sciences*, Paris, Ser. 2, 304: 853–856.

Barrick, R. E., W. J. Showers, and A. G. Fischer. 1996. Comparison of thermoregulation of four ornithischian dinosaurs and a varanid lizard from the Cretaceous Two Medicine Formation: Evidence from oxygen isotopes. *Palaios* 11: 295–305.

Bartholomew, G. A., and V. A. Tucker. 1964. Size, body temperature, thermal conductance, oxygen consumption, and heart rate in Australian varanid lizards. *Physiological Zoology* 37: 341–354.

Baynes, A. 1999. The absolutely last remake of *Beau Geste*: Yet another review of the Australian megafaunal radiocarbon dates. In A. Baynes and J. A. Long, eds., *Papers in Vertebrate Palaeontology*, p. 391. *Records of the Western Australian Museum*, Supplement, 57 (abstract).

Bierce, A. 1911. *The Devil's Dictionary*. Garden City, N.Y.: Doubleday and Company.

Bellwood, P. 1985. *Prehistory of the Indo-Malaysian Archipelago*. London: Academic Press. P. 370.

Bennett, A. F. 1973. Blood physiology and oxygen transport during activity in two lizards, *Varanus gouldii* and *Sauromalus hispidus*. *Comparative Biochemistry and Physiology* 46A: 673–690.

Bennett, D. 1998. *Monitor Lizards. Natural History, Biology and Husbandry*. Frankfurt am Main: Edition Chimaira. 352 pp.

Bilham, R., and P. England. 2001. Plateau "pop-up" in the great 1897 Assam earthquake. *Nature* 410: 806–809.

Blanchon, P., and J. Shaw. 1995. Reef drowning during the last deglaciation: Evidence for catastrophic sea-level rise and ice-sheet collapse. *Geology* 23: 4–8.

Blob, R. W. 2000. Interspecific scaling of the hindlimb skeleton in lizards, crocodilians, felids and canids: Does limb bone shape correlate with limb posture? *Journal of Zoology* 250: 507–531.

Borsuk-Białynicka, M. 1984. Anguimorphans and related lizards from the Late Cretaceous of the Gobi Desert, Mongolia. *Palaeontologia Polonica* 46: 5–105.

Bray, E. C. 1962. *A Million Years in Minnesota*. St. Paul: Science Museum of Minnesota. 49 pp.

Broecker, W. S., M. Ewing, and B. C. Heezen. 1960. Evidence for an abrupt change in climate close to 11,000 years ago. *American Journal of Science* 258: 429–448.

Brongersma, L. D. 1958. On an extinct species of *Varanus* (Reptilia, Sauria) from the island of Flores. *Zoologische Mededelingen, Rijksmuseum van Natuurlijke Historie* 36: 114–125.

Broom, R. 1898. On the affinities and habits of *Thylacoleo. Proceedings of the Linnean Society of New South Wales* 23: 57–74.

Brown, P. A., and J. P. Kennett. 1998. Megaflood erosion and meltwater plumbing changes during last North American deglaciation recorded in Gulf of Mexico sediments. *Geology* 26: 599–602.

Caldwell, M. W., R. L. Carroll, and H. Kaiser. 1995. The pectoral girdle and forelimb of *Carsosaurus marchesetti* (Aigialosauridae), with a preliminary phylogenetic analysis of mosasauroids and varanoids. *Journal of Vertebrate Paleontology,* 15: 516–531.

Camp, C. 1923. Classification of the lizards. *Bulletin of the American Museum of Natural History* 48: 289–481.

Campbell, K. E., Jr., C. D. Frailey, and J. Arellano L. 1985. The geology of the Rio Beni; Further evidence for Holocene flooding in Amazonia. *Contributions to Science, Natural History Museum of Los Angeles County* 364: 1–18.

Campbell, K. S. W., and M. W. Bell. 1977. A primitive amphibian from the Late Devonian of New South Wales. *Alcheringa* 1: 369–381.

Cane, M. A. 1998. A role for the tropical Pacific. *Science* 282: 59, 61.

Carrier, D. R. 1987. The evolution of locomotor stamina in tetrapods: Circumventing a mechanical constraint. *Paleobiology* 13: 326–341.

Carroll, R. L. 1988. Late Paleozoic and Early Mesozoic lepidosauromorphs and their relation to lizard ancestry. In R. Estes and G. Pregill, eds., *Phylogenetic Relationships of the Lizard Families*, pp. 99–118. Stanford: Stanford University Press.

Charles, C. 1997. Cool tropical punch of the ice ages. *Nature* 385: 681, 683.

Chorlton, W. 1983. *Ice Ages*. Amsterdam: Time-Life. 176 pp.

Clark, R. E. D. 1972. The Black Sea and Noah's flood. *Faith and Thought* 100: 174–179.

Cogger, H. G. 1983. *Reptiles and Amphibians of Australia*. Sydney: A. H. and A. W. Reed. 584 pp.

Collins, L. R. 1973. *Monotremes and Marsupials*. Washington, D.C.: Smithsonian Institution Press. 323 pp.

Covacevich, J. A., P. J. Couper, and C. James. 1993. A new skink, *Nangura spinosa* gen. et sp. nov., from a dry rainforest of southeastern Queensland. *Memoirs of the Queensland Museum* 34: 159–167.

Dana, J. D. 1885. *Manual of Geology*. 4th ed. New York: American Book. 1,088 pp.

Dansgaard, W., J. White, and S. Johnsen. 1989. The abrupt termination of the Younger Dryas climate event. *Nature* 339: 532–534.

Datan, I. 1993. Archaeological excavations at Gua Sireh (Serian) and Luvbang Angin (Gunung Mulu National Park), Sarawak, Malaysia. *Sarawak Museum Journal* 45: 1–192.

Dawson, T. J. 1983. *Monotremes and Marsupials: The Other Mammals*. London: Edward Arnold. 90 pp.

Dayton, L. 1994. The land where snakes are top dog. *New Scientist* 141, no. 1916: 14.

De Lisle, H. F. 1996. *The Natural History of Monitor Lizards*. Malabar, Fla.: Krieger Publishing. 201 pp.

Diamond, J. M. 1987. Did Komodo dragons evolve to eat pygmy elephants? *Nature* 326: 832.

Dodson, J., R. Fullagar, J. Furby, R. Jones, and I. Prosser. 1993. Humans and megafauna in a late Pleistocene environment from Cuddie Springs, north western New South Wales. *Archaeology in Oceania* 28: 94–99.

Edmund, A. G. 1962. Sequence and rate of tooth replacement in the Crocodilia. *Royal Ontario Museum; Life Sciences, Contribution* 56: 1–42.

———. 1969. Dentition. In C. Gans and T. S. Parsons, eds., *Biology of the Reptilia,* vol. 1: *Morphology A,* pp. 117–200. London: Academic Press.

Ericson, D. B., and G. Wollin. 1964. *The Deep and the Past.* New York: Grosset and Dunlap. 301 pp.

Erol, O., A. M. Sengör, A. Barka, and H. T. Genç. 1996. Was the Tarim Basin occupied entirely by a giant Late Pleistocene lake? *Abstracts with Programs of the 1996 Annual Meeting of the Geological Society of America* 28: A-497.

Estes, R. 1983. *Sauria terrestria, Amphisbaenia.* In *Handbuch der Palaeoherpetologie.* T. 10a. Stuttgart: Gustav Fischer Verlag. 249 pp.

———. 1984. Fish, amphibians and reptiles from the Etadunna Formation, Miocene of South Australia. *Australian Zoologist* 21: 335–343.

Estes, R., K. de Queiroz, and J. Gauthier. 1988. Phylogenetic relationships within Squamata. In R. Estes and G. Pregill, eds., *Phylogenetic Relationships of the Lizard Families,* pp. 119–281. Stanford, Calif.: Stanford University Press.

Etheridge, R., Jr. 1894. Official contributions to the palaeontology of South Australia. No. 6: Vertebrate remains from the Warburton or Diamantina River. *South Australia, Annual Report of the Government Geologist for the Year ended June 30th, 1894,* 19–22.

———. 1917. *Megalania prisca,* Owen and *Notiosaurus dentatus,* Owen: lacertilian dermal armour; opalised remains from Lightning Ridge. *Proceedings of the Royal Society of Victoria* 29: 127–133.

Evans, D. A., N. J. Beukes, and J. L. Kirschvink. 1997. Low-latitude glaciation in the Palaeoproterozoic era. *Nature* 386: 262–266.

Evans, S. E. 1994. A new anguimorph lizard from the Jurassic and Lower Cretaceous of England. *Palaeontology* 37: 33–49.

———. 1996. *Parviraptor* (Squamata: Anguimorpha) and other lizards from the Morrison Formation at Fruita, Colorado. In M. Morales, ed., *The Continental Jurassic,* pp. 243–248. *Museum of Northern Arizona Bulletin* 60.

Evans, S. E., G. V. R. Prasad, and B. K. Manhas. 2002. Fossil lizards from the Jurassic Kota Formation of India. *Journal of Vertebrate Paleontology* 22: 299–312.

Farlow, J. O. 1993. On the rareness of big, fierce animals: Speculations about body sizes, population densities, and geographic ranges of predatory mammals and large carnivorous dinosaurs. *American Journal of Science* 293-A: 167–199.

Farlow, J. O., D. L. Brinkman, W. L. Abler, and P. J. Currie. 1991. Size, shape, and serration density of theropod dinosaur lateral teeth. *Modern Geology* 16: 161–198.

de Fejervary, G. F. 1918. Contributions to a monograph on fossil Varanidae and on Megalanidae. *Annales Historico-Naturales Musei Nationalis Hungarici* 16: 341–465.

———. 1935. Further contributions to a monography of the Megalanidae and fossil Varanidae—with notes on recent varanians. *Annales Historico-Naturales Musei Nationalis Hungarici* 29: 1–465.

Flannery, T. F. 1994. *The Future Eaters.* Sydney: Reed. 423 pp.

Flower, W. H. 1868. On the affinities and probable habits of the extinct Australian marsupial *Thylacoleo carnifex* Owen. *Quarterly Journal of the Geological Society of London* 24: 307–319.

Frair, W., R. G. Ackman, and N. Mrosovsky. 1972. Body temperatures of *Dermochelys coriacea:* Warm turtle from cold water. *Science* 177: 791–793.

Fuller, S., P. Baverstock, and D. King. 1998. Biogeographic origins of goannas (Varanidae): A molecular perspective. *Molecular Phylogenetics and Evolution* 9: 294–307.

Furby, J. H., R. Fullagar, J. R. Dodson, and I. Prosser. 1993. The Cuddie Springs bone bed revisited. 1991. In M. A. Smith, M. Spriggs, and B. Fankhauser, eds., *Sahul in Review,* pp. 204–210. Canberra: Research School in Pacific Studies, Australian National University.

Gaffney, E. S. 1983. The cranial morphology of the extinct horned turtle, *Meiolania platyceps,* from the Pleistocene of Lord Howe Island, Australia. *Bulletin of the American Museum of Natural History* 175: 361–480.

Galloway, R. 1965. Late Quaternary climates in Australia. *Journal of Geology* 73: 603–618.

Galloway, R., and E. Kemp. 1984. Late Cainozoic environments in Australia. In M. Archer and G. C. Hickey, eds., *Vertebrate Zoogeography and Evolution in Australasia,* pp. 83–95. Perth: Hesperian Press.

Gaulke, M., and E. Curio. 2001. A new monitor lizard from Panay Island, Philippines. *Spixiana* 24:275–296.

Gilmore, C. W. 1928. The Fossil lizards of North America. *Memoirs of the National Academy of Sciences.* 2:1–197.

———. 1943. Fossil lizards of Mongolia. *Bulletin of the American Museum of Natural History* 81: 361–384.

Gilroy, R. 1995. *Mysterious Australia.* Mapleton, Queensland: Nexus Publishing. 287 pp.

Goudie, A. S. 1977. *Environmental Change.* Oxford: Oxford University Press. 244 pp.

Gradstein, F., and J. Ogg. 1996. A Phanerozoic time scale. *Episodes* 19: 3–6.

Green, B., D. King, M. Braysher, and A. Saim. 1991. Thermoregulation, water turnover and energetics of free-living Komodo dragons, *Varanus komodoensis. Comparative Biochemistry and Physiology* 99A: 97–101.

Gurnis, M., R. D. Mueller, and L. Moresi. 1998. Cretaceous vertical motion of Australia and the Australian-Antarctic discordance. *Science* 279: 1499–1504.

Guthrie, R. D. 1990. *Frozen Fauna of the Mammoth Steppe.* Chicago: University of Chicago Press. 323 pp.

Hecht, M. K. 1975. The morphology and relationships of the largest known terrestrial lizard, *Megalania prisca* Owen, from the Pleistocene of Australia. *Proceedings of the Royal Society of Victoria* 87: 239–250.

Hewitt, G. M. 1996. Some genetic consequences of ice ages, and their role in divergence and speciation. *Biological Journal of the Linnean Society* 58: 247–276.

Hoffstetter, R. 1969. Présence de Varanidae (Reptilia, Sauria) dans le Miocène de Catalogne. Considérations sur l'histoire de la famille. *Bulletin du Muséum National d'histoire Naturelle,* 40:1051–1064.

Hooijer, D. 1972. *Varanus* (Reptilia, Sauria) from the Pleistocene of Timor. *Zoologische Mededelingen* 47: 445–447.

von Huene, F. R. 1956. *Paläontologie und Phylogenie der Niederen Tetrapoden.* Jena: Gustav Fischer Verlag. 716 pp.

Huxley, T. H. 1887. Preliminary note on the fossil remains of a chelonian reptile, *Ceratochelys sthenurus,* from Lord Howe's Island, Australia. *Proceedings of the Royal Society of London* 42: 232–238.

Imbrie, J. I., and K. P. Imbrie. 1979. *Ice Ages.* Cambridge, Mass.: Harvard University Press. 224 pp.

Kennedy, K. A. R. 1989. Skeletal markers of occupational stress. In M. Y. Iscan and K. A. R. Kennedy, eds., *Reconstruction of Life from the Skeleton,* pp. 129–160. New York: Alan R. Liss.

Kennett, J. P., K. G. Cannariato, I. L. Hendy, and R. J. Behl. 2000. Carbon isotopic evidence for methane hydrate instability during Quaternary interstadials. *Science* 288: 128–133.

Kennett, J. P., and R. C. Thunell. 1975. Global increase in Quaternary explosive volcanism. *Science* 187: 497–503.

Kerr, R. A. 1994. How high was ice age ice? A rebounding earth may tell. *Science* 265: 189.

———. 2000a. A victim of the Black Sea flood found. *Science* 289: 2021.

———. 2000b. Gas blast for the dinosaurs? *Science* 287: 576–577.

King, D., and B. Green. 1993. *Goanna: The Biology of Varanid Lizards.* Sydney: New South Wales University Press. 102 pp. (The authors are listed in incorrect order on the front cover.)

Klicka, J., and R. M. Zink. 1997. The importance of recent ice ages in speciation: A failed paradigm. *Science* 277: 1666–1669.

Kordikova, E. G., P. D. Polly, V. A. Alifanov, Z. Rocek, G. F. Gunnell, and A. O. Averianov. 2001. Small vertebrates from the Late Cretaceous and Early Tertiary of the northeastern Aral Sea region, Kazakhstan. *Journal of Paleontology* 75: 390–400.

Kurtén, B. 1971. *The Age of Mammals.* New York: Columbia University Press. 250 pp.

———. 1972. *The Ice Age.* London: Rupert Hart-Davis. 179 pp.

Kyle, D. A. 1976. *A Pictorial History of Science Fiction.* London: Hamlyn. 173 pp.

Lamb, S., and D. Sington. 1998. *Earth Story.* London: BBC Books. 240 pp.

Lange, R. T. 1982. Australian Tertiary vegetation: Evidence and interpretation. In J. M. B. Smith, ed., *A History of Australasian Vegetation,* 44–89. Sydney: McGraw-Hill.

Lawver, L. A., L. M. Gahagan, and M. F. Coffin. 1992. The development of paleoseaways around Antarctica. *Antarctic Research Series* 56: 7–30.

Lee, M. S. Y. 1996. Possible affinities between *Varanus giganteus* and *Megalania prisca. Memoirs of the Queensland Museum* 39: 232.

———. 1997. Phylogenetic relationships among fossil and living platynotan squamates. *Philosophical Transactions of the Royal Society of London,* Ser. B, 352: 53–91.

Levesque, A. J., L. C. Cwynar, and I. R. Walker. 1997. Exceptionally steep north-south gradients in lake temperatures during the last deglaciation. *Nature* 385: 423–426.

Lewin, R. 1984. Fragile forests implied by Pleistocene pollen. *Science* 226: 36–37.

Losos, J. B., and H. W. Greene. 1988. Ecological and evolutionary implications of diet in monitor lizards. *Biological Journal of the Linnean Society* 35: 379–407.

Lydekker, R. 1886. Indian Tertiary and post-Tertiary Vertebrata. Parts 7 and 8: Siwalik Crocodilia, Lacertilia, and Ophidia, and Tertiary fishes. *Memoirs of the Geological Survey of India, Palaeontologia Indica,* Ser. 10, 3: 1–32.

———. 1888. *Catalogue of the Fossil Reptilia and Amphibia in the Brit-*

ish Museum (Natural History), Cromwell Road, S. W. Part I. The
orders Ornithosauria, Crocodilia, Dinosauria, Squamata, Rhyncho-
cephalia and Protorosauria. London: British Museum (Natural History). 309 pp.

———. 1889. Catalogue of the Fossil Reptilia and Amphibia in the British
Museum (Natural History), Cromwell Road, S. W. Part III. The order
Chelonia. London: British Museum (Natural History). 239 pp.

Main, A. R. 1978. Ecophysiology: Towards an understanding of Late
Pleistocene marsupial extinction. In D. Walker and J. C. Guppy, eds.
Biology and Quaternary Environments, pp. 169–183. Canberra: Aus-
tralian Academy of Sciences.

McMahon, T. A. 1973. Size and shape in biology. Science 179: 1201–1204.

McMahon, T. A., and J. T. Bonner. 1983. On Size and Life. New York:
Scientific American Library. 255 pp.

Meltzer, D. J., and J. I. Mead. 1985. Dating Late Pleistocene extinctions:
Theoretical issues, analytical bias, and substantive results. In J. I.
Mead and D. J. Meltzer, eds., Environments and Extinctions: Man
in Late Glacial North America, pp. 145–173. Orono: University of
Maine at Orono.

Millard, R. W., and K. Johansen. 1974. Ventricular outflow dynamics in
the lizard, Varanus niloticus: Responses to hypoxia, hypercarbia and
diving. Journal of Experimental Biology 60: 871–880.

Miller, G. H., J. W. Magee, B. J. Johnson, M. L. Fogel, N. A. Spooner, M.
T. McCulloch, and L. K. Ayliffe. 1999. Pleistocene extinction of
Genyornis newtoni: Human impact on Australian megafauna. Science
283: 205–208.

Molnar, R. E. 1990. New cranial elements of a giant varanid from Queensland. Memoirs of the Queensland Museum 29: 437–444.

Moore, P. D. 1998. Did forests survive the cold in a hotspot? Nature 391:
124–125, 127.

Morwood, M. J., F. Aziz, P. O'Sullivan, Nasruddin, D. R. Hobbs, and A.
Raza. 1999. Archaeological and palaeontological research in central
Flores, east Indonesia: Results of fieldwork 1997–98. Antiquity 73:
273–286.

Munn, H. W. 1974. Merlin's Ring. New York: Random House. 366 pp.

Murray, P. 1984. Extinction downunder: A bestiary of extinct Australian
Late Pleistocene monotremes and marsupials. In P. S. Martin and R. G.
Klein, eds., Quaternary Extinctions, pp. 600–628. Tucson: University
of Arizona Press.

Norell, M. A., and K. Gao. 1997. Braincase and phylogenetic relationships
of Estesia mongoliensis from the Late Cretaceous of the Gobi Desert
and the recognition of a new clade of lizards. American Museum
Novitates 3211: 1–25.

Norell, M. A., M. McKenna, and M. Novacek. 1992. Estesia mongoliensis,
a new fossil varanoid from the Late Cretaceous Barun Goyot Formation of Mongolia. American Museum Novitates 3045: 1–24.

Nydam, R. L. 2000. A new taxon of helodermatid-like lizard from the
Albian-Cenomanian of Utah. Journal of Vertebrate Paleontology 20:
285–294.

van Oosterzee, P. 1997. Where Worlds Collide. Melbourne: Reed Books.
234 pp.

Oppenheimer, S. 1998. Eden in the East. London: Weidenfeld and Nicol-
son. 560 pp.

Owen, R. 1854. On some fossil reptilian and mammalian remains from the Purbecks. *Quarterly Journal of the Geological Society of London* 10: 420–422.

———. 1859. Description of some remains of a gigantic land-lizard (*Megalania prisca*, Owen) from Australia. *Philosophical Transactions of the Royal Society of London* 149: 43–48.

———. 1871. On the fossil mammals of Australia, 4: Dentition and mandible of *Thylacoleo carnifex* with remarks on the arguments for herbivory. *Philosophical Transactions of the Royal Society of London* 161: 213–266.

———. 1880. Description of some remains of a gigantic land-lizard (*Megalania prisca* Owen) from Australia: Part II. *Philosophical Transactions of the Royal Society of London* 171: 1037–1050.

———. 1884a. *A History of British Fossil Reptiles.* 4 vols. London: Cassell and Company. 856 pp.

———. 1884b. Evidence of a large extinct lizard (*Notiosaurus dentatus,* Owen) from Pleistocene deposits, New South Wales, Australia. *Philosophical Transactions of the Royal Society of London* 175: 249–251.

———. 1886. Description of fossil remains, including foot-bones, of *Megalania prisca:* Part IV. *Philosophical Transactions of the Royal Society of London* 179: 327–330.

———. 1888. On parts of the skeleton of *Meiolania platyceps* (Ow.). *Philosophical Transactions of the Royal Society of London* 179: 181–191.

Owerkowicz, T., C. G. Farmer, J. W. Hicks, and E. L. Brainerd. 1999. Contribution of gular pumping to lung ventilation in monitor lizards. *Science* 284: 1661–1663.

Paladino, F. V., M. P. O'Connor, and J. R. Spotila. 1990. Metabolism of leatherback turtles, gigantothermy, and thermoregulation of dinosaurs. *Nature* 334: 858–860.

Parker, P. 1977. An ecological comparison of marsupial and placental patterns of reproduction. In B. Stonehouse and D. Gilmore, eds., *The Biology of the Marsupials,* pp. 273–286. London: Macmillan Press.

Paterson, D. 1993. Did Tibet cool the world? *New Scientist* 139, no. 1880: 29–33.

Paul, G. S. 1988. *Predatory Dinosaurs of the World.* New York: Simon and Schuster. 464 pp.

———. 1995. Super-sauropods, the greatest creatures on earth: How they lived, and how they got so big. *Kyōryūgaku Saizensen* 9: 70–85. (Text in Japanese.)

———. 1998. Terramegathermy and Cope's Rule in the land of titans. *Modern Geology* 21: 179–217.

Phillips, J. A. 1995. Movement patterns and density of *Varanus albigularis*. *Journal of Herpetology* 29: 407–416.

Pianka, E. R. 1994. Comparative ecology of *Varanus* in the Great Victoria Desert. *Australian Journal of Ecology* 19: 395–408.

———. 1995. Evolution of body size: Varanid lizards as a model system. *American Naturalist* 146: 398–414.

Pielou, E. C. 1991. *After the Ice Age.* Chicago: University of Chicago Press. 366 pp.

Pledge, N. S. 1990. The upper fossil fauna of the Henschke Fossil Cave, Naracoorte, South Australia. In S. Turner, R. A. Thulborn, and R. E. Molnar, eds., *Proceedings of the de Vis Symposium,* pp. 247–262. *Memoirs of the Queensland Museum* 28.

———. 1992. The Curramulka local fauna: A Late Tertiary fossil assemblage from Yorke Peninsula, South Australia. In P. F. Murray and D. Megirian, eds., *Proceedings of the 1991 Conference on Australasian Vertebrate Evolution, Palaeontology and Systematics,* pp. 115–142. Alice Springs, *The Beagle 9.*

Polyak, L., M. H. Edwards, B. J. Coakley, and M. Jakobsson. 2001. Ice shelves in the Pleistocene Arctic Ocean inferred from glaciogenic deep-sea bedforms. *Nature* 410: 453–457.

Pough, F. H. 1973. Lizard energetics and diet. *Ecology* 54: 837–844.

Pregill, G. K. 1989. Prehistoric extinction of giant iguanas in Tonga. *Copeia* 1989: 505–508.

Pregill, G. K., J. A. Gauthier, and H. W. Greene. 1986. The evolution of helodermatid squamates, with description of a new taxon and an overview of the Varanoidea. *Transactions of the San Diego Society of Natural History* 21: 167–202.

Raven, H. C. 1935. Wallace's line and the distribution of Indo-Australian mammals. *Bulletin of the American Museum of Natural History* 68: 179–293.

Reed, W. M. 1930. *The Earth for Sam.* New York: Harcourt, Brace and Co. 390 pp.

Rich, T. H., and B. Hall. 1979. Rebuilding a giant. *Australian Natural History* 19: 310–313.

Ride, W. D. L., P. A. Pridmore, R. E. Barwick, R. T. Wells, and R. D. Heady. 1997. Towards a biology of *Propleopus oscillans* (Marsupialia: Propleopinae, Hypsiprymnodontidae). *Proceedings of the Linnean Society of New South Wales* 117: 243–328.

Roberts, R. G., T. F. Flannery, L. K. Ayliffe, H. Yoshida, J. M. Olley, G. J. Prideaux, G. M. Laslett, A. Baynes, M. A. Smith, R. Jones, and B. L. Smith. 2001. New ages for the last Australian megafauna: Continent-wide extinction about 46,000 years ago. *Science* 292: 1888–1892.

Romer, A. S. 1956. *The Osteology of the Reptiles.* Chicago: University of Chicago Press. 772 pp.

Ross, C. A., and W. E. Magnusson. 1989. Living crocodilians. In L. Dow and C. Craig, eds., *Crocodiles and Alligators,* pp. 58–73. Sydney: Weldon Owen.

Rossignol-Strick, M., and N. Planchais. 1998. Climate patterns revealed by pollen and oxygen isotope records of a Tyrrhenian Sea core. *Nature* 342: 413–416.

Rothwell, R. G., J. Thomson, and G. Kaehler. 1998. Low-sea-level emplacement of a very large Late Pleistocene 'megaturbidite' in the western Mediterranean Sea. *Nature* 392: 377–380.

Ryan, W., and W. Pitman. 1998. *Noah's Flood.* New York: Simon and Schuster. 319 pp.

Sadlier, R. A. 1989. A review of the scincoid lizards of New Caledonia. *Records of the Australian Museum* 39: 1–66.

Salisbury, S. W., P. M. A. Willis, J. D. Scanlon, and B. Mackness. 1995. Plio-Pleistocene gigantism in *Quinkana* (Crocodyloidea; Mekosuchinae). In M. Augee, ed., *Quaternary Symposium,* p. 10. Sydney: Dept. of Zoology, University of New South Wales (abstract).

Scanlon, J. 1999. Letter to the editor. *Nature Australia* 26: 54.

Schaller, G. B. 1972. *The Serengeti Lion.* Chicago: University of Chicago Press. 480 pp.

Seebacher, F., G. C. Grigg, and L. A. Beard. 1999. Crocodiles as dinosaurs: Behavioural thermoregulation in very large ectotherms leads to high and stable body temperatures. *Journal of Experimental Biology* 202: 77–86.

Shackleton, N. J. 2000. The 100,000-year ice-age cycle identified and found to lag temperature, carbon dioxide, and orbital eccentricity. *Science* 289: 1897–1902.

Sher, A. V. 1996. Late-Quaternary extinction of large mammals in northern Eurasia: A new look at the Siberian contribution. In B. Huntley, W. Cramer, A. V. Morgan, H. C. Prentice, and J. R. M. Allen, eds., *Past and Future Rapid Environmental Changes,* pp. 319–339. Springer: Berlin.

Simpson, G. G. 1932. The supposed association of dinosaurs with mammals of Tertiary type in Patagonia. *American Museum Novitates* 566: 1–21.

———. 1977. Too many lines: The limit of the Oriental and Australian zoogeographic regions. *Proceedings of the American Philosophical Society* 121:107–120.

Singh, G. 1982. Environmental upheaval: Vegetation of Australasia during the Quaternary. In J. M. B. Smith, ed., *A History of Australasian Vegetation,* pp. 90–108. Sydney: McGraw-Hill.

Smith, M. J. 1976. Small fossil vertebrates from Victoria Cave, Naracoorte, South Australia. IV: Reptiles. *Transactions of the Royal Society of South Australia* 100: 39–51.

Smith, S. 1981. The Tasmanian tiger—1980. National Parks and Wildlife Service, Tasmania, Technical Report, 81/1:1–133.

Smith, W. R. 1930. *Aborigine.* London: Random House UK. 355 pp. (1996 edition).

Sondaar, P. Y., G. D. van der Bergh, B. Mubroto, F. Aziz, J. de Vos, and U. L. Batu. 1994. Middle Pleistocene faunal turnover and colonization of Flores (Indonesia) by *Homo erectus. Comptes Rendus de l'Academie des Sciences, Paris,* Ser. 2, 319: 1255–1262.

Spotila, J., M. P. O'Connor, P. Dodson, and F. V. Paladino. 1991. Hot and cold running dinosaurs: Body size, metabolism and migration. *Modern Geology* 16: 203–227.

Stafford, T. W., Jr., H. A. Semken Jr., R. W. Graham, W. F. Klippel, A. Markova, A. Smirnov, and J. Southon. 1999. First accelerator mass spectrometry [14]C dates documenting contemporaneity of nonanalog species in late Pleistocene mammal communities. *Geology* 27: 903–906.

Stebbins, G. L. 1950. *Variation and Evolution in Plants.* New York: Columbia University Press. 643 pp.

Steel, R. 1997. *Living Dragons.* London: Blandford. 160 pp.

Stocker, T. F. 1998. The seesaw effect. *Science* 282: 61–62.

Stokstad, E. 2001. Tooth theory revises history of mammals. *Science* 291: 26.

Struck, U., A. V. Altenbach, M. Gaulke, and F. Glaw. 2002. Tracing the diet of the monitor lizard *Varanus mabitang* by stable isotope analyses ($\delta^{15}N, \delta^{13}C$). *Naturwissenschaften* 89:470–473.

Surahya, S. 1989. *Atlas Komodo.* Yogyakarta: Gadjah Mada University Press.

Suzuki, N., K. Igarashi, and T. Hamada. 1991. Bipedal standing facility of *Varanus komodoensis* (Reptilia) in a wild habitat. *Scientific Papers of the College of Arts and Sciences, The University of Tokyo* 41: 95–105.

Tedford, R. H. 1994. Succession of Pliocene through medial Pleistocene mammal faunas of southeastern Australia. *Records of the South Australian Museum* 27: 79–93.

Tedford, R. H., R. T. Wells, and S. F. Barghoorn. 1992. Tirari Formation and contained faunas, Pliocene of the Lake Eyre Basin, South Australia. In P. F. Murray and D. Megirian, eds., *Proceedings of the 1991 Conference on Australasian Vertebrate Evolution, Palaeontology and Systematics,* pp. 173–193. Alice Springs, *The Beagle 9.*

Thompson, G. G., and E. R. Pianka. 2001. Allometry of clutch and neonate sizes in monitor lizards (Varanidae: *Varanus*). *Copeia* 2001:443–458.

Thompson, G. G., and P. C. Withers. 1997. Comparative morphology of Western Australian varanid lizard (Squamata: Varanidae). *Journal of Morphology* 233:127–152.

Thulborn, R. A. 1986. Early Triassic tetrapod faunas of southeastern Gondwana. *Alcheringa* 10: 297–313.

Thulborn, R. A., A. A. Warren, S. Turner, and T. Hamley. 1996. Early Carboniferous tetrapods in Australia. *Nature* 381: 777–779.

Tooley, M. J. 1989. Floodwaters mark sudden rise. *Nature* 342: 20–21.

Turner, F. B., R. I. Jennrich, and J. D. Weintraub. 1969. Home ranges and body size of lizards. *Ecology* 50: 1076–1080.

Ukraintseva, V. V. 1993. *Vegetation Cover and Environment of the "Mammoth Epoch" in Siberia.* Hot Springs: The Mammoth Site of Hot Springs, South Dakota, Inc.

Velichko, A. A. 1984. Late Pleistocene spatial paleoclimatic reconstructions. In A. A. Velichko, H. E. Wright Jr., and C. W. Barnosky, eds., *Late Quaternary Environments of the Soviet Union,* pp. 261–285. Minneapolis: University of Minnesota Press.

Velichko, A. A., L. L. Isayeva, V. M. Makeyev, G. G. Matsihov, and M. A. Faustova. 1984. Late Pleistocene glaciation of the Arctic shelf, and the reconstruction of Eurasian ice sheets. In A. A. Velichko, H. E. Wright Jr., and C. W. Barnosky, eds., *Late Quaternary Environments of the Soviet Union,* pp. 35–41. Minneapolis: University of Minnesota Press.

de Vis, C. W. 1883. On tooth-marked bones of extinct marsupials. *Proceedings of the Linnean Society of New South Wales* 8: 187–190.

———. 1885. On bones and teeth of a large extinct lizard. *Proceedings of the Royal Society of Queensland* 2: 25–32.

———. 1889. On *Megalania* and its allies. *Proceedings of the Royal Society of Queensland* 6: 93–99.

———. 1900. A further trace of an extinct lizard. *Annals of the Queensland Museum* 5: 6.

Walker, E. P. 1975. *Mammals of the World.* 2 vols. 3rd ed. Baltimore: Johns Hopkins University Press. 1500 pp.

Warren, A. A., and T. Black. 1985. A new rhytidosteid (Amphibia, Labyrinthodontia) from the Early Triassic Arcadia Formation of Queensland, Australia, and the relationships of Triassic temnospondyls. *Journal of Vertebrate Paleontology* 5: 303–327.

Warren, A. A., and M. N. Hutchinson. 1983. The last labyrinthodont? A new brachyopoid (Amphibia, Temnospondyli) from the Early Jurassic Evergreen Formation of Queensland, Australia. *Philosophical Transactions of the Royal Society of London,* Ser. B, 303: 1–62.

Warren, A. A., T. H. Rich, and P. Vickers-Rich. 1997. The last last labyrinthodonts? *Palaeontographica* (A), 247: 1–24.

Webb, G., and C. Manolis. 1989. *Crocodiles of Australia.* Sydney: Reed Books. 158 pp.

Webb, G., H. Heatwole, and J. de Bavay, 1971. Comparative cardiac anatomy of the Reptilia. l. The chambers and septa of the varanid ventricle. *Journal of Morphology* 134: 335–350.

Wells, R., D. R. Horton, and P. Rogers. 1982. *Thylacoleo carnifex* Owen

(Thylacoleonidae): Marsupial carnivore? In M. Archer, ed., *Carnivorous Marsupials*, pp. 573–585. Sydney: Royal Zoological Society of New South Wales.

White, J. W. C., and E. J. Steig. 1998. Timing is everything in a game of two hemispheres. *Nature* 394: 717–718.

Wiffen, J. 1991. *Valley of the Dragons*. Auckland, New Zealand: Random Century. 128 pp.

Wilkinson, J. E. 1995. Fossil record of a varanid from the Darling Downs, southeastern Queensland. *Memoirs of the Queensland Museum* 38: 92.

Willis, K. J., A. Kleczkowski, K. M. Briggs, and C. A. Gilligan. 1999. The role of sub-Milankovitch climatic forcing in the initiation of the Northern Hemisphere glaciation. *Science* 285: 568–571.

Willis, P. M. A., and R. E. Molnar. 1997a. A review of the Plio-Pleistocene crocodilian genus *Pallimnarchus*. *Proceedings of the Linnean Society of New South Wales* 117: 224–242.

Willis, P. M. A., and R. E. Molnar. 1997b. Identification of large reptilian teeth from Plio-Pleistocene deposits of Australia. *Journal and Proceedings of the Royal Society of New South Wales* 130: 79–92.

Woillard, G. 1979. Abrupt end of the last interglacial s.s. in north-east France. *Nature* 281: 558–562.

Woodard, G. D., and L. F. Marcus. 1973. Rancho La Brea fossil deposits: A re-evaluation from stratigraphic and geological evidence. *Journal of Paleontology* 47: 54–69.

Woodward, A. S. 1888. Note on the extinct reptilian genera *Megalania*, Owen, and *Meiolania*, Owen. *Annals and Magazine of Natural History* (Ser. 6), 1: 85–89.

Worthy, T. H., A. J. Anderson, and R. E. Molnar. 1999. Megafaunal expression in a land without mammals: The first fossil faunas from terrestrial deposits in Fiji (Vertebrata: Amphibia, Reptilia, Aves). *Senckenbergiana biologica* 79: 237–242.

Wroe, S. 1999. Killer kangaroos and other murderous marsupials. *Scientific American* 280, no. 5: 58–64.

Wroe, S., T. J. Myers, R. T. Wells, and A. Gillespie. 1999. Estimating the weight of the Pleistocene marsupial lion, *Thylacoleo carnifex* (Thylacoleonidae: Marsupialia): Implications for the ecomorphology of a marsupial super-predator and hypotheses of the impoverishment of the Australian marsupial carnivore faunas. *Australian Journal of Zoology* 47: 489–498.

Yadagiri, P. 1986. Lower Jurassic lower vertebrates from Kota Formation, Pranhita-Godavari Valley, India. *Journal of the Palaeontological Society of India* 31: 89–96.

Zahavi, A., A. Zahavi, N. Zahavi-Ely, and M. P. Ely. 1997. *The Handicap Principle*. Oxford: Oxford University Press. 286 pp.

Zietz, A. 1899. Notes upon some fossil reptilian remains from the Warburton River, near Lake Eyre. *Transactions, Proceedings and Reports of the Royal Society of South Australia* 23: 208–210.

# Index

RALPH E. MOLNAR
was for many years Section Leader for Vertebrate and Invertebrate Paleontology and Geology, and Senior Curator of Vertebrate Paleontology, at Queensland Museum in Australia. Now retired in Arizona, he continues to pursue his interests in paleontology and his hobbies, including raising carnivorous plants and investigating sightings of UFOs, yowies, and abominable snowpersons.

Milton Keynes UK
Ingram Content Group UK Ltd.
UKHW050322290923
429575UK00003B/45

9 780253 343741